The Methodology of Experimental Economics

The experimental approach is a driving force behind some of the most exciting developments in economics. The "experimental revolution" was based on a series of bold philosophical premises that have remained until now mostly unexplored. This book provides the first comprehensive analysis and critical discussion of the methodology of experimental economics, written by a philosopher of science with expertise in the field. It outlines the fundamental principles of experimental inference in order to investigate their power, scope, and limitations. The author demonstrates that experimental economists have a lot to gain by discussing openly the philosophical principles that guide their work, and that philosophers of science have a lot to learn from the ingenious techniques devised by experimenters in order to tackle difficult scientific problems.

Francesco Guala is Senior Lecturer in Philosophy, Department of Sociology and Philosophy, University of Exeter, United Kingdom. He has also worked at Imperial College London and the University of Cergy-Pontoise, France, and serves as an External Research Associate of the Center for Philosophy of the Natural and Social Sciences, London School of Economics. Dr. Guala has published extensively in philosophy and social science journals, including *Philosophy of Science*, *Studies in the History and Philosophy of Science*, *Economics and Philosophy*, *Journal of Economic Methodology*, *Experimental Economics*, *Journal of Theoretical Politics*, and *Behavioral and Brain Sciences*. He won the INEM (International Network for Economic Methodology) Prize in 2001 for best article on the methodology of economics written by a young scholar and, in the same year, won the History of Economic Analysis Award for best article on the history of economic thought written by a young scholar.

The Methodology of Experimental Economics

FRANCESCO GUALA
University of Exeter

CAMBRIDGE
UNIVERSITY PRESS

CAMBRIDGE UNIVERSITY PRESS
Cambridge, New York, Melbourne, Madrid, Cape Town, Singapore,
São Paulo, Delhi, Dubai, Tokyo

Cambridge University Press
32 Avenue of the Americas, New York, NY 10013-2473, USA

www.cambridge.org
Information on this title: www.cambridge.org/9780521618618

First published 2005

A catalog record for this publication is available from the British Library

Library of Congress Cataloging in Publication data
Guala, Francesco, 1970–
The methodology of experimental economics / Francesco Guala.
p.   cm.
Includes bibliographical references and index.
ISBN 0-521-85340-0 (hardcover) – ISBN 0-521-61861-4 (pbk.)
1. Economics – Methodology.   2. Economics – Experiments – Methodology.   I. Title.
HB131.G83   2005
330′.072′4 – dc22   2004031079

ISBN 978-0-521-85340-8 Hardback
ISBN 978-0-521-61861-8 Paperback

Transferred to digital printing 2009

To Francesca

Contents

Analytical Table of Contents

1 Introduction

This book aims to show that methodology is important and useful for experimental economists, but also that philosophers of science can learn from experimental economics. It is neither a handbook nor a textbook of experimental economics.

Part one: inferences within the experiment

2 Inside the Laboratory

Experimental economists often complain that replication is not valued enough in their discipline, but they fail to notice a crucial distinction between mere repetition and replication. In this chapter, I introduce experimental economics to the novice by describing the replication of an experimental phenomenon known as the "decay of overcontribution" in public goods games. Particularly important is the role of pilots and the extensive checking for errors performed before, during, and after the experiment.

3 Hypothesis Testing

The Hypothetico-Deductive (HD) model is a very popular, very simple, and very general model of scientific method. It can be used to highlight some basic logical problems of testing, such as the Duhem-Quine problem: no hypothesis can be logically falsified by the empirical evidence. As a consequence, scientific reasoning must include a logic of inductive inference. In this chapter, I also show what kind of hypotheses are routinely tested by scientists, and introduce an important distinction between "data" and "phenomena".

4 Causation and Experimental Control

The key to experimental control is the controlled variation of one variable keeping the other (background) conditions fixed. The rationale of variation can be explained using a second important model of scientific method, the perfectly controlled experimental design. This model is particularly important in experiments aimed at testing causal hypotheses. Causes can be used to control or manipulate their effects. Causal relations can be deterministic or probabilistic, and the perfectly controlled experiment exemplifies a situation in which the statistical association between variables reflects the underlying causal relations.

5 Prediction

Laboratory experimentation helps to tackle the Duhem-Quine problem constructively, or to draw tight inductive inferences from the evidence to a given hypothesis. Much philosophical literature, however, has focused on the wrong aspects of this inductive step, by stressing the importance of predictive success. In fact, the crucial advantage of the experimental method is that it allows the control of the background assumptions upon which strong inductive inferences rest. This thesis is illustrated using the example of preference reversal experiments.

6 Elimination

Bayesian confirmation theory stresses the importance of the background, but for the wrong reasons. Scientists' prior beliefs should not be given too much weight in confirmation theory. What matters is whether the background factors have been controlled by means of an effective experimental design. The experimental method is best characterized as a procedure of eliminative induction, in which factors that may potentially disturb the inference from the evidence to a hypothesis are checked one by one, until all sources of error have been controlled for. Experiments on preference reversals provide several examples of this strategy at work.

Part two: inferences from the experiment

7 External Validity

There is a trade-off between the internal validity of an experimental result (whether a given laboratory phenomenon or mechanism has been correctly identified) and its external validity (whether the results can be generalized from the laboratory to the outside world). External validity is a genuine problem and cannot be solved by metaphysical speculation or methodological stipulation. It is an issue that must be tackled and solved empirically.

8 Economic Engineering

The best example of successful external validity inference is provided by cases of economic engineering, in which a piece of the real world is shaped so as to mirror the conditions of a laboratory experiment. I illustrate this procedure using the early auctions of the Federal Communication Commission as an example. The key external validity step is taken by comparing field evidence with experimental evidence and using a so-called no-miracle argument.

9 From the Laboratory to the Outside World

"Radical localists" argue that experimental results only apply to laboratory circumstances, or to real-world circumstances that have been engineered so as to resemble the lab. In reality, when experimenters cannot shape the real world so as to fit the laboratory, they can try to shape the laboratory so as to mimic the target system in the real world. Winner's curse experiments illustrate this principle at work. The inference from experiment to the real world is a special kind of analogical argument, in which the inference is strengthened by making sure that the two systems are similar in all relevant (causal) respects.

10 Experiments as Mediators

Models and experiments share several important characteristics. Both are systems that are created to aid scientists in their investigations of a target system. They are "mediating tools," an intermediary step in the process connecting our theoretical speculations with the real world. Like models, experiments can be closer to abstract theory or to application. The purpose of an experiment is often to test the robustness of a phenomenon rather than its applicability to a particular real-world situation.

Acknowledgments

This project started a decade ago at King's College London, was developed during my Ph.D. years at the London School of Economics, and was concluded at the University of Exeter. During this period of time, I have received financial support from various sources; I should mention especially a TMR ("Marie Curie") scholarship of the European Union, the European TMR Network FMRX CT 96005, the center THEMA at the University of Cergy–Pontoise, a research fellowship of the Cognitive Science Laboratory of the University of Trento, and the research leave scheme of the University of Exeter.

I am indebted to a number of people, starting with Mary Morgan and Philippe Mongin, who first encouraged me to study economic experiments as a young graduate student and then in many ways supported this project throughout the years. Philippe became my Ph.D. supervisor jointly with John Worrall; they gave me an example of intellectual rigor that I have always tried to imitate and probably never attained. My first close encounter with experimental economics was via the experimental group of the University of East Anglia. Chris Starmer, Bob Sugden, and Robin Cubitt, in particular, have always been willing to discuss openly and frankly the methodological reasoning that informs their research. I must also thank Roberto Burlando for having guided me into the practical aspects of experimental work, and Luigi Mittone for several discussions on methodological matters. Parts of Chapter 10 derive from a working paper Luigi and I coauthored a few years ago.

Other people who have contributed with several useful comments and critiques over the years include Paul Anand, Roger Backhouse, Barry Barnes, Mark Blaug, John Broome, Nancy Cartwright, Marco Del

Seta, Dan Hausman, Peter Lipton, Gian Maria Martini, Deborah Mayo, Phil Mirowski, Edward Nik-Kah, Nigel Pleasants, Andrea Salanti, Joep Sonnemans, and Sang Wook Yi. I am particularly grateful to John Dupré, Dan Hausman, Mary Morgan, Matteo Motterlini, and the anonymous readers who have read and criticized preliminary versions of the manuscript. Parts of some chapters have been presented at conferences and seminars around the world. I should in particular thank the audiences at the universities of Cambridge, Bergamo, East Anglia, LSE, Paris, Pisa, Rotterdam, Siena, Tilburg, and Trento as well as those at the 1998 HES conference in Montreal, the 1999 Theorizing and Experimentation conference in Rovaniemi, the 1999 SILFS conference in Ferrara, the Model-Based Reasoning conferences held in Pavia (1998 and 2000), the PSA 2002 meeting in Milwaukee, the workshop on the methodology of experimental economics held in Nottingham (2003), the 2004 INEM meeting in Amsterdam, and the performativity of economics workshop in Paris (2004).

The views defended in this book are closest to those of Nancy Cartwright, Dan Hausman, and John Dupré – which is quite natural, I suppose, given that Nancy and Dan have been my teachers at the LSE, and John is now a senior colleague of mine at Exeter. I am sure that in several places I have failed to fully recognize my debt toward their ideas. This initial acknowledgment is offered by way of an apology. Of course, I am entirely responsible for all the mistakes that remain.

Finally, bits and pieces of various published articles are scattered throughout the book: Chapters 5 and 6 build upon "Artefacts in Experimental Economics: Preference Reversals and the Becker-DeGroot-Marschak Mechanism" (*Economics and Philosophy* 16, 2000, pp. 47–75). Traces of "The Problem of External Validity (Or 'Parallelism') in Experimental Economics" (*Social Science Information* 38, 1999, pp. 555–73) can be detected in Chapter 7, and are reprinted by permission of Sage. Chapter 8 is a revised version of "Building Economic Machines: The FCC Auctions" (*Studies in History and Philosophy of Science* 32, 2001, pp. 453–77), reprinted by permission of Elsevier Science. Chapter 9 reproduces material from "Experimental Localism and External Validity" (*Philosophy of Science* 70, 2003, pp. 1195–205), reprinted by permission of the Philosophy of Science Association. "Experiments as Mediators in the Non-laboratory Sciences" (*Philosophica* 62, 1998, pp. 901–18) is the starting point for much of Chapters 9 and 10. Parts of "Models, Simulations, and Experiments" (in *Model–based Reasoning: Science, Technology, Values*, ed. by L. Magnani and N. J. Nersessian, pp. 59–74) are also reproduced in Chapter 10 by permission of Kluwer.

ONE

Introduction

On October 9, 2002, the news began to circulate that Daniel Kahneman and Vernon Smith had been awarded the Bank of Sweden Prize in Economic Sciences in Memory of Alfred Nobel. The prize was not entirely unexpected. For a few years, the names of prominent experimental economists had been in the list of plausible Nobel candidates, and everybody agreed that it was just a matter of time. Yet, for the community of experimenters, the event was epoch making: laboratory work was recognized officially as one of the most important advancements in the last half century of social science.

This wasn't the first time that the Nobel Prize had been assigned to the proponents of doctrines or approaches that do not enjoy universal acceptance within the profession. And economists like Maurice Allais, Herbert Simon, and Rheinhard Selten, who had contributed in many ways to the birth and development of experimental economics, already figured among the laureates. But Simon, Allais, and Selten had been prized for their work in other areas of economic theory, and the 2002 award constituted an innovation in at least two major respects. First, it recognized the work of a scholar who according to the conventions of contemporary academia should not be labeled as an "economist." Daniel Kahnemann was prized "for having integrated insights from psychological research into economic science, especially concerning human judgment and decision making under uncertainty" (Nobel Press Release 2002). This was a contribution "from without," by a prominent psychologist who challenged many key ideas in the mainstream economic tradition.

But secondly, and more importantly perhaps, the other half of the prize was devoted to recognizing a *methodological innovation*, rather than a

1

contribution to the body of economic theory. Vernon Smith was prized "for having established laboratory experiments as a tool in empirical economic analysis, especially in the study of alternative market mechanisms" (ibid.). Of course methodological innovations carry important novel theoretical insights with them: if you look at the world with different instruments, you are likely to notice different things. And in fact the work of Vernon Smith includes important theoretical contributions as well. But the Nobel committee was keen to stress that

Economics has been widely considered a non-experimental science, relying on observation of real-world economies rather than controlled laboratory experiments. Nowadays, however, a growing body of research is devoted to modifying and testing basic economic assumptions; moreover, economic research relies increasingly on data collected in the lab rather than in the field. (ibid.)

It was this change in the nature of economic science that was primarily recognized by means of the 2002 award.

Why experiment in economics?

Until fairly recently, most economists believed that controlled experimentation had little to offer economic science. These beliefs are voiced in some of the most influential methodological writings of the last couple of centuries; despite their different views about what constituted "good" methodological practice, everybody seemed to agree that economic research was bound to take place mostly outside the laboratory. In 1836, John Stuart Mill claimed that "there is a property common to almost all the moral sciences, and by which they are distinguished from many of the physical; that is, that it is seldom in our power to make experiments in them." About a century later, Lionel Robbins wrote that "our belief [in economic generalizations] does not rest on the results of controlled experiments." And Milton Friedman in his influential essay on the methodology of positive economics also states that "we can seldom test particular predictions in the social sciences by experiments explicitly designed to eliminate what are judged to be the most important disturbing influences." Such statements – partly because of the prestige of their authors, partly because they reflected the views of the average economist – migrated in the most popular economics textbooks, upon which generations of economists have been trained. "Economics must be a non-laboratory science," wrote Richard Lipsey, given that "it is rarely, if ever, possible to conduct controlled experiments with the economy."

Samuelson and Nordhaus similarly claim that "economists [. . .] cannot perform the controlled experiments of chemists or biologists because they cannot easily control other important factors," and even in the *Encyclopaedia Britannica* one reads that "there is no laboratory in which economists can test their hypotheses."[1]

Nowadays these claims have become obsolete. Economists perform hundreds of laboratory experiments every year, and routinely test their theories in the laboratory. But *why* do they do such things? Why is experimentation highly considered today, but was not half a century ago? An obvious answer is that the success of experimental economics was made possible by several profound changes in the discipline of economics as a whole. In order to write a proper history of experimental economics (something that still has to be done), one would certainly have to look back at the birth of expected utility and game theory in the forties and fifties, examine the rise and fall of general equilibrium analysis in the sixties and seventies, discuss the high expectations and frustrations that accompanied the development of econometrics, and probably much else.[2]

However, this is not supposed to be a book of history, and I shall leave it to someone else to do a proper historical job. When I ask, Why experiment in economics? I am not concerned with the reasons why experimental economics is more popular today than, say, one hundred years ago. That is an interesting question indeed, but it is not the question of this book. This book asks, Why experiment in economics? *in general*, or as a matter of principle. I take the latter to be an ahistorical question, in the sense that it asks what sort of knowledge social scientists can collect in the laboratory, regardless of time, place, and context. It is, in other words, an *epistemological* question about the capacity of laboratory experimentation to produce knowledge about economic matters. As such, it does not investigate the

---

[1] Cf. Mill (1836, p. 124) and Robbins (1932, p. 74). The Friedman and Lipsey quotes are taken from Starmer (1999, p. 1), Samuelson and Nordhaus' from Friedman and Sunder (1994, p. 1), the *Encyclopaedia Britannica* from Davis and Holt (1993, p. 4, n. 2).

[2] Brief reconstructions of the early history of experimental economics can be found in Smith (1991a; 1992), Davis and Holt (1993), Friedman and Sunder (1994), Roth (1995), and Hargreaves Heap and Varoufakis (1995). Leonard (1994) is the only contribution by a professional historian, but focuses on bargaining experiments only. Mirowski (2002) vividly reconstructs the milieu of mid-twentieth-century economics, in which the conditions for the birth of experimental economics were created, and devotes a short section to Vernon Smith's research program (pp. 545–51). Two Ph.D. dissertations at Notre Dame are beginning to explore the history of experimental economics in more depth (Lee unpublished, Nik-Kah unpublished).

contingent factors that as a matter of fact have prompted economists to turn to the laboratory. It investigates the reasons why economists *should* (or should not) endorse the experimental approach, by articulating what experiments can (and cannot) do for them.

Why experiment?

The experimental method is as old as science itself, and became the hallmark of the most successful science – physics – during the scientific revolution of the Renaissance. One may think, therefore, that the basic elements of the experimental method must be well understood by now. Surprisingly, this is not the case. Of course the literature on experiments is large. All great scientists since Galileo have put forward their own views about the proper use of experiments, and professional philosophers have added more thoughts on this topic. Philosophers' views, however, were for a long time detached from experimental science as it is practiced in real-world laboratories. They typically followed from fairly abstract speculations about the nature and sources of knowledge in general, as if laboratory experimentation had no peculiar features of its own that justified a separate analysis. And curiously, scientists tended to follow philosophers on this track, by privileging philosophical speculation instead of reflecting on their *real* practice. (Perhaps this is partly because many great scientists of the seventeenth and eighteenth centuries were also great philosophers, with very precise views on abstract epistemological matters.)

A fundamental assumption of this book is that in contrast, philosophy of science must look closely at the messy business of science in the making. There has been a movement in the past two decades toward a philosophy of science that is more sensitive to the details of scientific practice. This body of work has provided an impressive amount of data about the methods actually followed by scientists in their everyday work, and has been an important source of inspiration for what is to follow.[3] There are two good reasons, however, not to start from an account of "*the* experimental method" as it emerges from these studies. First of all, there is no such account. The picture emerging from such studies is patchy, and to try to distill a unified story out of this material would be an unlikely task. Secondly, and related to this, this material is diverse because scientific

---

[3] For a survey of the so-called new experimentalist movement in the philosophy of science, cf. Ackerman (1989), Franklin (1998), or Morrison (1998a).

practice probably *is* diverse. Experimental physics or biochemistry (which have been studied extensively) may follow in part different procedures from experimental economics. We should not expect these disciplines to be *entirely* different, of course, but neither should we presuppose that they are identical. An account of experimental science based on physics or biology may not fit the bill of the experimental economist.

Why economics?

Experimental economics seems a pretty odd topic for a philosopher of science. Most philosophers interested in normative questions about science (How should genuine scientific knowledge be generated? How do scientists avoid falling into error? What exactly is scientific knowledge *knowledge of*? and so on) tend to look at the natural sciences, because these are supposed to be the most advanced disciplines with respect to both their results and their methodology. However, as I said, we should not assume that what works for physics or biochemistry should work for economics too. After all, the methods of discovery and validation that scientists use must be right for the particular domain or sort of thing they are studying.

So one reason to look at economics is that it might teach philosophers of science something new. Indeed, in the second part of this book, I shall argue that natural science-based methodology tends to neglect an important problem of scientific inference: the problem of external validity, or how to generalize experimental results to nonlaboratory settings. I shall suggest that in this case, natural scientists have something to learn from social scientists, rather than the other way around. Another reason to focus on economics is that laboratory experimentation is a fairly new methodology there, and the field has not crystallized yet on a set of rules of "good" scientific practice. As opposed to physics or chemistry, in which methodological discussion has arguably had little effect in changing scientists' habits, the social sciences seem to be more permeable to philosophical arguments. Within experimental economics, in particular, methodological discussion is alive and well, and also potentially influential in the way in which the discipline is taking shape.

Many philosophers of science suffer from a sense of guilt of being useless, and every now and then make an attempt to write something aimed at helping scientists in their everyday work. Unfortunately, their way of engaging scientists is to start with a painfully analytical discussion of abstract issues that are only very indirectly of practical relevance. In this book, I have tried to avoid that approach and to stick close to the real concerns of experimental scientists. I have tried to keep the detours in abstract philosophical arguments under control, and put philosophy at use in understanding the rationale of down-to-earth methodological principles.

What is in this book?

So this is a book of methodology, devoted to a relatively small but growing field of the social sciences called experimental economics. It discusses the techniques used by experimenters in order to investigate economic phenomena, evaluates them, and occasionally puts forward some advice about how to revise our thinking about laboratory experimentation (its goals, its role, and its tools). It is divided in two parts. In the first part (Inferences within the Experiment), I discuss how the experimental method allows the drawing of tight inferences from data to phenomena, and from phenomena to their causes within a given experimental setting. In the second part (Inferences from the Experiment), I show how, and under which circumstances, it is sometimes possible to generalize an experimental result from laboratory circumstances to some real-world situation.

Eight themes recur in the chapters that follow:

1. Experimental and theoretical knowledge often grow independently from each other.
2. The growth of experimental knowledge is slow and piecemeal: experimental scientists learn little by little rather than attempting great leaps forward.
3. "Local" knowledge of the experimental systems and the background conditions in which hypotheses are tested is crucial for the reliability of scientific inference.
4. Experimental inference is based on "eliminative induction," a process aimed at eliminating alternative interpretations of empirical evidence.
5. Social practices and conventions play a key role in determining when a given phenomenon or hypothesis is accepted as established by experimental means.
6. Experiments act as "mediators" between the real world and the theories, models, and hypotheses we devise to explain its functioning.

7. Real-world applicability is nonetheless the ultimate aim of science, which conveys knowledge of causal relations for intervention and policy making.
8. It is difficult to extend experimental results to real-world circumstances unless we are able to shape the experiment and the real world so as to resemble each other.

Some of these ideas are not new and have been defended before by philosophers working on experimental methodology. Others are philosophically less conventional and spring from the observation of the concrete problems economic experimenters face in their day-to-day work. They should sound familiar to experimenters, however, because they try to capture the concerns that drive their research. The accent on applicability and policy making, for instance, is a common feature of many recent overviews and discussions of experimental economics.[4] However, experimenters often make use of a *rhetoric* of scientific method that is far removed from the reality of their work. Partly this has to do with the fact that methodological norms sometimes serve the purpose of marking political alliances and contrapositions (e.g., "economics vs. psychology") independently of the similarity or dissimilarity of the methods that are effectively used. And partly, it is an effect of the fact that scientific method (like science itself) often evolves by imitation, and paradigmatic experiments often are much more effective in shaping the practices of a discipline than are explicit methodological pronouncements.[5] So, like Vernon Smith, I believe that by and large "if you look at what experimental economists do, not what they say, you get the right picture of science learning" (1994, p. 129). Ideally, one would like to capture all the "good" methodological practice implicit in the work of experimental economists, purged from the "bad" rhetoric that obfuscates experimenters' achievements and, sometimes, diverts them onto dangerous or dead-end trails. By doing this, one can do a service to philosophers and practitioners alike.

It may be useful also to clarify right from the start what this book is *not* about. This is not a handbook of experimental economics. There are already some excellent surveys of the main results in the field (Kagel and

---

[4] Cf. e.g., Plott (1987), Roth (1991, 2002), Ross Miller (2002), Smith (2002).
[5] A reader indicated Cox, Robertson, and Smith (1982); Isaac, McCue, and Plott (1985); and Grether and Plott (1979) as examples of seminal experiments on (respectively) auctions, public goods, and decision making that were also extremely influential in setting the methodological standards of the discipline. I will return to the issue of the economics–psychology divide again in Chapter 11.

Roth, eds., 1995; Plott and Smith, eds., in press) and it is not my purpose to add to these resources. In some chapters, I shall of course illustrate some experiments, but my aim is neither to be exhaustive nor to present the state of the art in the discipline. I also have no intention to add anything new to the experimental literature – I am interested in philosophy and methodology, and the experiments described in the book are always used as examples of methodological principles, never as novel contributions to experimental research.

This is not a textbook of experimental methodology either, nor a book that will teach you how to design experiments. If you are interested in that, you should consult Davis and Holt (1993), Friedman and Sunder (1994), Bergstrom and Miller (1997), or Friedman and Cassar (2004). This book falls somewhere between the concrete instructions of a textbook and the abstract analysis of classic philosophy monographs. The main reason to pitch the discussion at such a level is that concrete techniques of experimental design must fulfill the sort of higher-level requirements that are customarily discussed in the philosophy of science literature. However, unfortunately, it is difficult to reach any firm conclusion about higher-level methodological principles or requirements unless one keeps in mind the subject-specific problems experimental scientists have to deal with in their daily work. One purpose of this book is to try to fill the gap between abstract philosophy and concrete scientific practice.

Despite my efforts to simplify as much as possible, the level of detail may be at times a bit demanding for the noneconomist. However, I am afraid it is difficult to say anything meaningful about the method of science by sticking to totally unrealistic examples of scientific reasoning like "there are black swans in Australia, therefore it is not true that all swans are white."[6] To help the novice, I have provided a lengthy description of a "normal" economic experiment (Chapter 2) and, for the nonphilosophers, an introduction to basic notions of testing and confirmation in Chapters 3 and 4. Hopefully, both economists and philosophers of science will find something useful in this book.

As far as I'm concerned, I have certainly learned a lot by studying experimental economics. I have been surprised by the ingenious

[6] For the nonphilosophers: "all swans are white" is an alleged example of law-of-nature or scientific theory that is widely abused in the philosophical literature. The first to use the example was, as far as I know, John Stuart Mill (a great philosopher-economist, incidentally).

techniques invented by experimenters, I have been amazed by the robustness of certain results, and not the least, I have realized how addictive experimental work can be. It would be nice if I could transmit only part of this fascination to my nonscientist readers. And at the same time I hope the economists will be convinced that interesting and useful things are taught in (some) philosophy classes.

PART ONE

INFERENCES WITHIN THE EXPERIMENT

TWO

Inside the Laboratory

Before we get into the heart of the matter, it is worth putting on the table an example of experimental research in economics. I shall focus on an experiment that I know very well, because I was personally involved in it. Because the experiment as a whole is too complicated to be fully described here, I'll just focus on part of it. The part I shall look at is a typical example of a *replication*. Replicating someone else's results is not the sort of thing that will win you a Nobel Prize in economics. Fame and prestige derive from revolutionary results that affect the direction of research in the discipline. However, replications are not without importance and in fact constitute a large part of everyday work in experimental science.

Replications

A common reaction, when people first hear about experimental economics, is to say that *of course* people are not the sort of thing you can experiment with. Behind this reaction lies the thought that human beings are quite different from, say, atoms or molecules: they possess that elusive capacity that philosophers call "free will." So, the argument goes, their behavior does not obey laws, as does the behavior of physical entities. What's the point of experimenting then?

This argument takes several sophisticated forms in the social science literature, and even a cursory discussion would take away more space than required for my purposes. But the above worries can be easily dismissed: as any experimenter knows, *human behavior is highly predictable*. Two caveats are in order: first, it is true that the behavior of an individual x may be hard to predict exactly. But the behavior of aggregates (even relatively small groups) seems to follow very systematic patterns, in some circumstances. Here's the second caveat: the circumstances matter, and predictability often depends on the creation of rather precise choice-situations.

None of these qualifications, however, constitutes a fundamental difference with respect to experimentation in the natural sciences. The average behavior of an aggregate (of, say, particles) is always easier to predict than the behavior of each constituent. And outside well-specified initial and boundary conditions, the movement of physical particles may be as unpredictable as the choices of human beings.

The best proof of the predictability of human behavior is the fact that several behavioral patterns can be (and are) replicated at will by experimenters in their economic laboratories. However, experimental economists often complain that replications are not valued enough in their discipline (e.g., Smith 1994, p. 128; Rubinstein 2001, pp. 625–6). Given that the replication of a previously observed result normally will not be published in economic journals, it is argued that researchers lack the incentive to check the results reported by others.

It is necessary to distinguish, however, between the *replication* and the mere *repetition* of an experiment.[1] Repetition is the business of doing an experiment again, trying to keep exactly the same design as in the original. Repetitions are rarely performed, and only for the purpose of checking that the data provided by another experimenter are reliable and trustworthy – that data of *that* kind really follow from *that* particular experimental design. It is true that repetitions do not receive much recognition, unless, that is, one discovers some major flaw or fraud in an already published result. And in fact, experiments are repeated only when there is some serious doubt about other people's data (when the data appear "too strange to be true").

Genuine *replications* are different, in that they usually involve some (minor or major) modification of the original design. Another way to put it is that scientists usually aim at reproducing experimental *phenomena* rather than the experiments themselves. The difference will become clearer when I introduce the distinction between "data" and "phenomena" in Chapter 3. But roughly, a phenomenon (or "effect,"

---

[1] The distinction introduced here is well known in the philosophical literature on experiments, although other authors sometimes use another terminology (e.g., "replication" vs. "reproduction"). Cf. e.g., Cartwright (1991) and Radder (1996). Backhouse (1997, Ch. 11) applies it to econometrics.

as scientists sometimes call it) is a robust regularity lying "below" a set of data. The data typically have different characteristics, depending on the experimental setting or design used, but different data sets with different characteristics can be used to infer the existence of the same phenomenon. In this book, I discuss several well-known, widely replicated phenomena in experimental economics, including the decay of overcontribution in public goods experiments (this chapter), the preference reversal phenomenon (Chapters 5 and 6), and the winner's curse phenomenon (Chapter 9).

Scientists' everyday talk does not distinguish properly between replication and repetition. Indeed physicists have the tendency to use the term *replication* rather liberally (cf. Mulkay and Gilbert 1986), a habit that has probably misled scientists in younger and less-established disciplines to believe that physicists repeat experiments much more often than they actually do. However, once the distinction is made clear, it turns out that economists also do a lot of replications. Replications (the reproduction of a result by means of a slightly or radically different design) are an important part of everyday experimental work. But why? What is the purpose of a replication?

It is difficult to answer such a question at this stage, before I have introduced some basic methodological tools and concepts. However, a preliminary answer goes as follows: successful replications help in proving that an experimental phenomenon is *robust* to small or big changes in the experimental setup. In several respects, experimenters speak of robustness in the same way as theoretical scientists do – a theoretical result being "robust" when it does not depend on some detail of the situation or on the assumptions used to derive it. A good scientific result is always robust to some kind of variation or change, either in the concrete experimental setup or in the abstract initial conditions of a theoretical model.

Normal science

Of course, replications are not all equally interesting. Like many other replications, the example I discuss in this chapter does not have much interest per se, but played a role in a wider research project. I choose to discuss this replication mainly because of its simplicity but also because I am looking for an example that is representative of a large portion of research done every single day in economics laboratories around the world, which is neither path breaking nor epoch making. It is "normal science."

The notion of normal science was introduced by the historian Thomas Kuhn in his highly influential monograph, *The Structure of Scientific Revolutions* (1962), and has since become part of the vocabulary of the academia at large. Kuhn noticed that for long periods of time, scientific research within a discipline follows a rather continuous path, characterized by the accumulation of results that contribute to the growth of knowledge without challenging the basic tenets of the received view. Such periods of normal science (or "paradigmatic" science, in the sense that research is carried out within a dominant scientific paradigm) are separated by more or less abrupt scientific revolutions. The major presuppositions informing normal science research are questioned and rejected during a revolution. Typical examples are the Copernican revolution that turned astronomy upside down in the Renaissance, or the Quantum revolution of the early twentieth century. In economics, plausible candidates are the marginalist revolution of the late nineteenth century and the Keynesian revolution half a century later.[2]

If you want to understand the methodological principles that govern research in a given discipline, you are better off focusing on a piece of normal science than on path-breaking results. Great scientific work is usually subversive of the status quo, at least to some extent, and therefore is not representative of what goes on within the paradigm. (Otherwise, it would not be surprising and impressive – anyone could have done it!) In the next chapters, to be sure, I discuss some results that are (or have been) regarded by the majority of practitioners as surprising or anomalous. But for the time being, we'd better look at something less controversial – which does not mean that we are dealing with work of no importance. As I argue later in the book, the bulk of scientific progress is constituted by the slow accumulation of "small" results rather than great revolutions.

One warning to the reader before I begin. If you are an expert experimental economist, most of what you will find here will be fairly familiar to you. In contrast, if you have never seen an economic experiment, it is worth reading this chapter quite carefully. It will tell you what the whole book is about, and probably will also help you get rid of a few myths and prejudices about scientific research.

---

[2] On revolutions in economics, cf. the essays in Latsis (ed. 1976).

Public goods experiments

If you want to become an experimental economist, where do you start? The obvious answer is that most novel research stems from previous scientific research. The myth of scientific genius, the legend of Newton watching a falling apple and suddenly conceiving of the theory of universal gravitation is precisely that: just a myth. In reality, science is characterized by continuity. Each discovery follows from some previous empirical result, a previous theory, or a puzzle (an inconsistency, an anomaly) highlighted by previous research. As a consequence, you cannot do science unless you know at least the recent history of your discipline. Training at graduate level is supposed to give you just that: knowledge of the background, of the theories and models that are generally accepted by the scientific community, of the problems that remain to be solved, of the most challenging empirical results that call for more research efforts. Even in a fairly young discipline like experimental economics, this background knowledge can be vast. There are hundreds of people working in the area, and constantly producing new results. Moreover, one must also keep an eye on the relevant economic theory, which also progresses (more or less independently) as empirical knowledge accumulates.

Public goods experiments are among the most widely replicated in economics and experimental psychology, sociology, and political science (where they are also known as "social dilemma" experiments). Hundreds, if not thousands, of experiments of this sort have been performed in the last three decades, and a complete review of the literature would be almost impossible.[3] (In this sense, a public goods experiment is genuine "normal science.") Public goods experiments have a reputation for being particularly delicate. John Ledyard uses the analogy of a physics experiment on free fall down a plane using a table-tennis ball instead of a steel ball. In theory, that is, if the experiment is designed appropriately, there should be no difference between the two objects. However, in practice, minor imperfections in the design (e.g., a small breeze or some friction on the plane) will provoke nonnegligible differences. "Public goods and dilemma experiments are like using table-tennis balls; sensitive enough to be really informative but only with adequate control" (Ledyard 1995, p. 115). Given this sensitivity to minor details of the design, they constitute a good challenge for the novice.

---

[3] The best point of entry is Ledyard's (1995) review article in the *Handbook of Experimental Economics*.

Table 2.1. *A Prisoner's Dilemma Game*

		Other	
		Defect	Cooperate
You	Defect	(5,5)	(10,1)
	Cooperate	(1,10)	(8,8)

A so-called public good has two essential characteristics: it is (a) *nonri-valled* and (b) *nonexcludable*. This means that once it has been produced, (a) many people can consume it at the same time and (b) you cannot make individuals pay for what they consume. There are several examples of public goods, the most commonly cited being clean air, public parks, public health service, national security, public scientific research, and so on. There are also "public bads," such as crime, pollution, and global warming. Public goods suffer from a fundamental problem: although they are beneficial to everybody, it is in the interest of each individual to free ride and not contribute to their production. If everybody else travels by bicycle, it is in my interest to travel by car: my using the car will affect the quality of the air only slightly, and therefore, I shall have the best of both worlds – fairly clean air and quick transportation. However, if everybody reasons this way, we shall all use our cars, pollute the air, and create massive traffic jams.

We can start by representing the situation in terms of a so-called pris-oner's dilemma game: suppose there are just two players, you and another person. Suppose also there are just two possible actions: either cooperate (taking the bike, in the example above) or defect (taking the car). The structure of the game is represented in Table 2.1.

In theory, the specific numbers in the table are not important, as long as your payoffs satisfy the following structure: defect/cooperate (you defect, the other cooperates) > cooperate/cooperate > defect/defect > cooper-ate/defect. Notice that *given the other player's move*, "defect" always gen-erates a higher payoff than "cooperate." If the other defects, it is better to defect too; if the other cooperates, it is still better to defect. In game-theoretic jargon, "defect" is a *dominant strategy*. But then if all players play the dominant strategy, the outcome (5, 5) will be inferior to what could have been achieved by both cooperating (8, 8). The moral is that individually rational action seems to lead to a disappointing social result (a "Pareto-inferior" outcome, in technical terms).

Let us now complicate the game by increasing the number of players, and by allowing more strategies than just "cooperate" and "defect." Here's a classic public goods experimental environment: imagine there are four players, each with an endowment of twenty points or tokens. The tokens can be either spent for one's own private leisure, or invested in the production of the public good (say, clean air). In the example above, this corresponds qualitatively to the decisions to travel by car or by bike, respectively. The difference is that in the public goods environment, one can split her own endowment in many different ways between private and public investment (zero and twenty, one and nineteen, two and eighteen, and so on), and must take into account the decisions of three other players. All players play simultaneously and anonymously – at the moment of taking her decision, each subject ignores the identity of the other subjects in her group, and how much they are contributing. Now, imagine the total sum invested in the production of the public good by all players is multiplied by a factor of two and then divided equally among the players, independently of the amount of their individual contribution (remember: the good is nonexcludable). For groups of four players, the payoff function of each player is:

$$p_i = 20 - g_i + 0.5 \sum_{j=1}^{4} g_j,$$

where 20 is the total number of tokens to be shared between a "private" $(20 - g)$ and a "public" account (g). The parameter 0.5 is called the "production factor" and specifies how much of the public good is enjoyed by each individual, for each unit invested by the group as a whole. This particular environment is characterized by a linear relationship between total payoff and contribution to the public project, and complete symmetry among players (the payoff function is identical for everybody). The linear environment has been widely used because of its simple structure, which makes it very intuitive and easily understandable. There are many possible variations on this baseline situation, which I shall not explore here for reasons of space (but see Ledyard 1995).

So what happens in the linear environment? According to standard economic theory, the public good should not be produced, that is, there should be no contributions to the public project. This conclusion is reached by assuming that each player is selfish, only cares about money, and is perfectly rational in the sense of Nash rationality. A Nash equilibrium is such that the strategy implemented by each player is the best move given the strategies of the other players: in equilibrium, no player has an incentive to change her own strategy, in other words. In our case, it is clear that the best move – regardless of what the others do – is not to contribute anything. If the others do not contribute anything, why should one give her own tokens, given that she would get back only half of each token contributed to the project? If the others do contribute one token, it is still best not to contribute anything, and enjoy the fruits of the others' contribution plus one's own full endowment. And so on: this reasoning can be iterated for all levels of contribution, and the moral will always be the same.

The linear environment is easy to understand also because the social optimum (everybody contributes all their tokens) and the Nash equilibrium (no one contributes anything) are at the extreme boundaries of the contribution range (zero to twenty). By manipulating the payoff function, it is possible to create situations in which the Nash equilibrium lies in the middle of the range, but for the time being, we can ignore such subtleties.[4] Notice that the prediction of standard economic theory has some important policy implications. It suggests, for instance, that the production of public goods cannot be delegated to individuals or to the market. Clean air, for instance, must be preserved by a system of controls and punishments (fines for those who pollute, prohibitions from traveling by car) or by means of incentives that change the payoff structure of the situation and effectively transform it into a different game altogether. Similarly, projects like cancer research should be heavily subsidized by the government, rather than relying on individual donations. But the problem is, are we really sure that economic theory's prediction is correct?

People, after all, do contribute voluntarily to cancer research and other public goods; in many circumstances, they seem to abstain voluntarily from polluting the environment, and so on. These are "stylized facts," which can be further corroborated by means of quantitative data collected in the field. To know how many people contribute to charities and by how much, however, may not satisfy our scientific curiosity. In general, we would also like to know *why* they do so. Such knowledge, in turn, could be used to inform new policies in order to create, for example, better conditions for the production of public goods. Surveys are one possible approach to the study of motives and individual attitudes: we could, for example, interview a sample of people who donate to the

---

[4] See Keser (1996) and Isaac and Walker (1998).

Red Cross and a sample of people who do not, and see whether we can identify the different attitudes behind their behavior. However, this may still be insufficient. Surveys do not allow the observation of how people *actually* behave in various different situations. In principle, one could ask, What would you do if the situation were so and so? – for example, How much would you contribute to this public good if the government stopped subsidizing it? – but economists have been traditionally suspicious of this sort of questionnaire. How do we know that respondents will tell the truth? They may be deluding themselves, or reasoning on the basis of unrealistic assumptions about the relevant circumstances (about other people's behavior or motivations, for instance). It is much better, if possible, to observe what people do, instead of asking what they *would* do.

Here experiments can be of great help. The general idea behind experiments is that we can put real people in circumstances that have been accurately prepared by the experimenters, and then observe their behavior. In the case of public goods, the earliest experiments go back to the seventies, and right from the beginning, generated results that are in stark contrast with the predictions of economic theory.

Overcontribution and decay

In a standard one-shot linear public goods experiment it is common to observe an average level of contribution of about fifty percent of the individual endowment, instead of nothing as predicted by theory. A substantial portion of subjects are willing to contribute something to the public project. If you let them play the game more than once, however, giving them constant feedback about the payoffs and the average contribution levels in previous rounds, subjects' behavior seems to change. The relatively high initial levels of contribution tend to diminish over time, converging toward the Nash equilibrium. These two phenomena are sometimes referred to in the literature as "overcontribution" and "decay."[5] Even long series of repetitions, however, are insufficient to eliminate contributions completely. Figure 2.1 reproduces the pattern

---

[5] Notice that according to standard game theory, *finitely* repeated public goods games should be distinguished sharply from *infinitely* repeated ones. An infinitely or indefinitely repeated public goods game has an infinite number of equilibria, including cooperative ones; using a backward induction argument, it is possible to show that in contrast, free riding is the only rational strategy in a finitely repeated environment (i.e., one-shot and finitely repeated games are identical in this respect). In this chapter, I confine my discussion to the finitely repeated environment only.

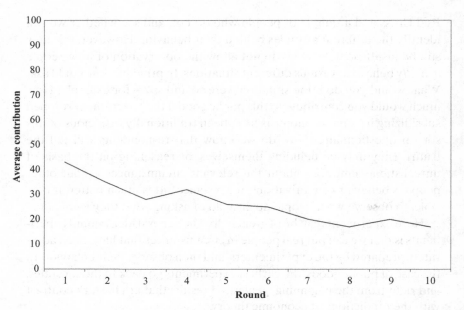

Figure 2.1. The overcontribution and decay effect (from Isaac, Walker, and Thomas 1984).

observed by Isaac, Walker, and Thomas (1984) in a seminal and widely cited experiment. Isaac et al. experimented with various production factors; the data represented here were generated using a factor of .3, leading to a drop of about 25 percent in the contribution level over a period of ten rounds. However, it has been established that after fifty rounds or more, there are still individuals who are willing to put something into the public project.

Overcontribution and decay call for an explanation, and various hypotheses have been proposed so far. One is that people are not entirely selfish. Their behavior may be dictated at least in part by altruistic motives, because, for example, they care about other people's payoffs.[6] An alternative explanation is cognitive in character. When you try to explain the logic of Nash rationality to nonexperts, especially with respect to prisoner's dilemma-like situations, the claim that it is rational to free ride usually generates some puzzlement. People seem to compare the Pareto-optimal solution with the Nash equilibrium and reason along these lines:

[6] Altruism may be of various kinds: for example, one may enjoy the act of giving in itself, or instead derive pleasure from the benefit procured to other individuals. James Andreoni (1995) has investigated experimentally these different forms of altruism.

in Pareto, we all make X points; in Nash we make Y points; X is greater than Y, therefore it is rational to play the Pareto strategy. If you have ever taught an introductory game theory course, you know that it is quite difficult to make students appreciate the strategic aspects of the situation. So, perhaps subjects at the beginning of the experiment do not fully appreciate what is in their best interest. After a few rounds, however, they might begin to "learn" about the game and form more adequate preferences and beliefs about the situation (which includes the behavior of other players, of course). Slowly, the outcome of the experiment converges toward the equilibrium.

There have been various attempts to model these two factors (altruism and learning) in theoretical terms. One idea is to add to the standard model of economic behavior an "altruism" parameter and an "error" component. Palfrey and Prisbey (1996, 1997), for example, assume that individuals follow a decision rule of the following kind:

- Contribute if $r/V < 1 + a + \varepsilon$;
- Do not contribute otherwise.

Here r is the value of a privately consumed token, V is the value to the individual of a token contributed to the public good, a is an altruism parameter, and ε is an error term. The altruism and the error coefficients can be estimated empirically on the basis of experimental data.[7] In the standard linear environment, if an individual is perfectly rational and selfish (that is, if her a and ε are equal to zero), the r/V ratio will always be greater than one. The prediction then is "no contribution" as in the standard game-theoretic analysis. An individual contributes, according to this model, only when the a and ε factors are strong enough to turn the balance.

A characteristic feature of models like this one, however, is that they assume individuals to be substantially uninterested in the choices of other people. This seems to be unrealistic. We know from everyday experience that many people value, for instance, fairness and equality, and are ready to cooperate as long as the other members of their group do the same. We shall call such people "reciprocators" or "conditional cooperators." Reciprocators are important because their existence suggests a natural explanation of the decay phenomenon. Imagine a population composed entirely of two types of players: free riders and altruists. Free riders contribute nothing in a public goods game, and altruists always contribute

---

[7] See also Anderson, Goeree, and Holt (1998) for a similar attempt.

something. No decay should be observed, unless some player realizes that she is not *really* an altruist or a free rider, or that the game was not as it initially appeared to be. Now imagine there is a third kind of player, the conditional cooperator. Such a player contributes only if the others do the same. Perhaps she will initially give something to the public project, but soon her efforts will be frustrated by the free riders in her group. Therefore, she will lower her contribution to put it in line with the group average, with the effect of further depressing the average itself. The decay phenomenon has now been triggered and perpetuates itself, making contributions spiral down quickly toward zero. (Perhaps a few "pure" altruists will remain who guarantee a low level of contributions above the Nash equilibrium.)

Reciprocating behavior is difficult to model, and despite some attempts in this direction, the results so far have not been entirely satisfactory.[8] The existence of conditional cooperators, however, can be established also by nontheoretical means, that is, by means of a purely experimental approach. The most relevant attempts in the literature are those by Offerman, Sonnemans, and Schram (1996), and by Fischbacher, Gächter, and Fehr (2001). Offerman and his collaborators use a psychological test known as the Strategy Method in order to identify players of different types, and then test the classification they have obtained against the behavior of the same subjects in a standard public goods game. Fischbacher et al., in contrast, classify using a method called the Decomposed Game technique, and again check for consistency using a one-shot public goods game.

These approaches are similar in that the subjects always play the public goods game in "heterogeneous" groups, that is, in "mixed" groups made up of cooperators, reciprocators, and free riders. An interesting question is what would happen if subjects of each type could play the game in "homogeneous" groups, that is, in groups made up of players of the same kind. The obvious hypothesis to test is whether different (homogenous) groups would display patterns of cooperation that depart significantly from the standard decay phenomenon. Jointly with Roberto Burlando, an experimenter from the University of Turin, I set out in the spring of 2002 to test this hypothesis experimentally. The results strongly confirm the "heterogeneous agents" hypothesis (cf. Burlando and Guala, in press), but to describe this experiment in full detail goes well beyond the purposes

---

[8] Cf. e.g., Sugden (1984), Rabin (1993), Falk and Fischbacher (2000); see also Fehr and Fischbacher (2002) for a general discussion.

of this chapter.[9] I instead focus on what *all* experiments on repeated public goods games have in common, namely the replication of the decay effect. If you want to demonstrate that a given effect is the result of one or more causal factors, you need to show that variations in that factor or condition result in deviations from the benchmark phenomenon. In experimental jargon, the benchmark phenomenon acts as a *control* (a contrast-case) for the effect elicited by means of the new design. As a consequence, the benchmark effect has to be replicated. In the next sections, I briefly describe how.

Preparing the experiment

What do you need in order to run an economic experiment? Two things are indispensable: some money and an economic laboratory. Money is necessary to pay the subjects who participate in the experiment. The habit of compensating subjects distinguishes experimental economics from other neighboring disciplines, such as experimental psychology.[10] Because economists often use university students in their experiments, the sums involved are not huge (a student's time is supposed to cost less than the time of, say, a doctor or a lawyer); in this case, Roberto and I could rely on a budget of about 2,000 euros. The second indispensable facility is the laboratory. Unlike physicists, economists do not need particularly complicated machines or instruments of observation. Most economic experiments can even be performed by paper and pencil, in which case all you need is a room big enough to seat the subjects for each session. However, nowadays many (perhaps most) experiments are run on computer networks. This has several advantages. First of all, the environment is standardized: all subjects receive the same information on the screen, and uniformity of circumstances is easily achieved. Secondly, the interaction among subjects is much quicker. In a paper-and-pencil public goods game, each player writes his or her own contribution on a sheet of paper; the sheet is collected by the experimenter, who then compares the contributions of all players in each group, calculates the payoffs, writes them on the same sheets together with other information of interest (group

---

[9] Other papers have since appeared that report similar results on agents' heterogeneity: cf. e.g., Burlando and Webley (1999) and Gachter and Thoni (2004). In experiments like these, subjects are "exogenously" selected and matched by the experimenters. Experiments with "endogenous" group formation include Ehrhart and Keser (1999) and Page, Putterman, and Unel (2002).

[10] In Chapter 11, I discuss the issue of monetary incentives in more depth.

average, etc.), and gives them back to the subjects. If you consider that an experiment must involve fifty subjects at least and a repeated public goods game can go on for sixty rounds, you can imagine how time consuming such a procedure is – not to mention the mistakes one is prone to make with such a method: it is fairly easy to miscalculate, to return the wrong sheet to the wrong player, and so on. A computerized network solves all these problems for you: the calculations are made instantly, feedback is quick, and computers do not get tired or make mistakes (well, mostly . . .).

Computers also have another advantage. The data are stored immediately and are available for data analysis almost straight away. You don't have to spend hours typing data into SPSS or some other statistical software. Of course all these advantages come at a cost. First of all, you need a computer room that you can use for your experiment, with a network of terminals linked to a central server. Many universities have these facilities nowadays, so this is not a huge obstacle. A bigger problem is that you need special software for each experiment you run, and obviously someone must create it. There are, to be sure, some standard packages available on the Internet, and others that experimenters are happy to circulate for free, if requested. But in 99 percent of the cases, your experimental design will be slightly different from that of your colleagues, and you will need at least to modify their software. This requires programming skills, which not all economists possess. In our case, we were lucky to rely on the expertise and apparatus of an established lab, the Computable and Experimental Economics Laboratory (CEEL) at the University of Trento, in the north of Italy. Many experiments are run at CEEL every single year. Not only were the logistic problems easily solved there (computer rooms, recruitment of subjects, etc.), but the software was tailor-made for us by the staff of the laboratory.

The experimental design: the instructions

There are three major aspects in the design of an economic experiment: the instructions, the software, and the physical environment. The instructions must contain all the information, and only the information, that the subjects need in order to perform the experimental task. Usually the instructions are printed on paper and distributed at the beginning of the experiment. Writing up the instructions is not a trivial matter, and in running a replication, one is advised to follow carefully the instructions of the benchmark experiment (instruction sets are circulated freely among experimenters).

Instructions must also be clear, sharp, and of the right length. The first two requirements do not need any justification: unless your explicit goal is to create confusion in the subjects, you had better seek simplicity. Some concepts require careful formulation: it may make some difference, for instance, to use the expression "private account" instead of "individual account"; in general, it is a good idea to avoid morally charged terms like "altruism," "egoism," and so on, which might induce subjects to believe that you expect them to behave in the "right" way. It is also a good idea not to use economists' jargon: first of all, most people do not know the meaning of economics' technical terms, and secondly, some of these terms are normatively laden. You usually do not want to tell subjects what is in their "rational" interest to do, or what the "equilibrium" of the game is, and so forth. You might not even want to let them know that there exists a "rational" solution to the game. On the other hand, you should not make the opposite mistake of being simplistic. The instructions must not be too short, and it is important that the subjects understand all the subtleties of the situation they will face in the experiment.

Every replication departs in some respects from the design or designs that have been used by other researchers. Typically, some deviations are not grounded in theory. For example: how many rounds should a finitely repeated public goods game go on for? Standard theory does not say, for any number of rounds should provide the same result (a contribution of zero throughout the game). Although the theory does not seem to capture what happens in experiments, common experimental wisdom suggests that indeed the number of rounds does not matter – the decay of contribution has similar characteristics regardless of the number of rounds played. Secondly, what kind of production coefficient should we use? Does it make a difference whether it is set at .5 rather than .4, for example? We chose .5 because we thought it would have been easier for our subjects to understand the mechanism of production if we used this coefficient (in order to calculate it, you just need to multiply by two and divide by four the total amount of tokens in the public account, something even the most mathematically illiterate subjects should be able to do). According to standard theory, once again, changes in the production function should not matter at all (as long as it is linear). However, there is evidence from past experiments that increments in the production function can affect behavior: as the value of the tokens invested in the public project increases, people seem to be willing to invest more in it.[11]

Another decision concerned the level of incentives. Practical considerations play an important role in this case. On the one hand, there is an obvious trade-off between the level of payoffs and the number of experimental subjects, given a fixed budget allocated to the experiment. Secondly, the administration of our university did not allow us to pay each student more than a certain amount of money for reasons that have to do with Italian tax legislation.[12] For these reasons, we chose payoffs aiming at an average individual earning of 11 euros, for a task of less than one hour. This is in line with the incentives used in the majority of economic experiments. (In theory, lower incentives may produce more "noise," that is, marginally more subjects who provide confused or unreliable responses. Notice, however, that the sums involved are more than what an average Italian student can hope to earn in a part-time job.)

The software

The software for the experiment was written by Marco Tecilla at CEEL, starting from an early draft of our instructions. I will not go into the details of the process of writing the software, which is obviously heavily dependent on the sort of programming tools that you use. It is, however, worth spending a few words on the basic structure of the final product. Before launching the software, the experimenter must specify the fundamental parameters of the experiment, such as the number of players, the number of groups, the endowment, the payoffs, and so on. Then, the application must be started on each computer terminal. The software automatically assigns a number to each player/terminal and randomly matches the terminals to form the experimental groups: for example, terminal 1 with terminals 4, 7, and 8; terminal 2 with 3, 5, and 6; and so on. (We played with groups of four.) The first window to appear on the screen when the players have read the instructions includes a request to write one's name and university ID. The game starts when all "OK" buttons have been clicked. The first window asks how much each player is willing to contribute to the public account. Once all players have specified their

---

[11] Cf. Isaac, Walker, and Thomas (1984), Kim and Walker (1984).

[12] In principle, we could have paid them above the threshold, but this would have created such a bureaucratic mess that we decided not to do it.

contributions, a new window appears with the average contribution level of the group in the last round, the individual payoff in the last round, and the accumulated earnings. The new window also asks the player to specify the contribution for the second round, and so on. Three rounds are played for training, with no "real" payoffs. At the end of the training period, subjects are informed that the rounds played for real are about to start. After twenty rounds, the total payoff will appear on the screen of each subject. These payoffs are still expressed in experimental tokens, and will be converted in real money (euros, in this case) according to a prespecified exchange rate.

The physical environment

The physical environment is where the experiment takes place. A university computer room, as I said, is the standard environment for experimental economics. Before you actually run an experiment, it is a good idea to have a look at the place and see whether it fits your requirements. There are trivial issues to be addressed (e.g., Is it big enough? Can we seat all the people we need?), but also slightly more sophisticated ones: Are the subjects too close to each other? Can we isolate them from one another if we want to? Are the terminals reliable, are they too slow, do they freeze or break down too often? If the experiment requires that subjects receive some public feedback (e.g., if it is an auction or a market experiment), you might also need a projector or a blackboard to post prices and record transactions. (Obviously, the screen or board must be visible from all terminals.) It is a good idea to take extra sets of instructions, some paper, and extra pens for the subjects if they are allowed to take notes during the experiment. In our case, we also had to use some cardboard partitions in order to isolate subjects from one another. (This is necessary to implement anonymity: the players must not know who they are playing with or what the other group members are doing.) If you want to play "transparent" lotteries, you might also need a set of dice, some numbered balls in a bowl, or whatever else can be used for a random draw. Finally, you need sheets to record the earnings of each subject at the end of the experiment; these should be signed by the experimenter and countersigned by the subject as a receipt. (I shall bracket here the amazing bureaucratic complications that certain universities and funding bodies manage to impose on such a relatively simple procedure. Paying subjects can be as annoying as the administration wants it to be!)

Pilots, bugs, and checking

I have now described the main ingredients and the apparatus needed for a typical economic experiment. When all this is ready, you might want to run the experiment itself – and this would be a mistake. Experimental science is, in many respects, similar to engineering.[13] As any engineer will tell you, no new technology works the first time it is used. It may work approximately, or mostly, but in general several tests must be performed before a machine can be used safely and efficiently. The same happens in experimental science. Like a complicated machine, an experiment requires lots of checking and testing before it is run for real. One major concern of course is money: you don't want to blow your whole budget at once. You want to be sure that the experiment will run smoothly, that it will generate data, and that the data will give you exactly what you need in order to answer the questions that motivated the experiment.

This may seem a minor detail, but it is not. In fact, several hours are routinely spent pretesting an experiment, just to check that everything is fine. The crucial tool in this phase is the so-called pilot experiment. A pilot is basically the experiment itself, but run on a smaller scale and under careful monitoring. Sometimes the experimenters themselves play the role of experimental subjects in the pilot, with help from friends, research students, and colleagues. The payoffs are not paid at the end of the pilot, and the data are not used to test the hypotheses at stake; but in all other respects, the pilot should closely resemble the real experiment. All pilot subjects are usually provided with a notebook in order to record the problems they encounter. The rules of the game may be applied less rigidly than in the real experiment – communication between players may be allowed, for instance. Sometimes the pilot will be interrupted in order to clarify some aspect of the experiment, or to correct some mistake. Just to give you an idea of how important pilots are, here's a list of some errors, bugs, and imperfections we found in our experimental design and apparatus. The pilot was run in the same computer room where the real experiment would take place. We acted as "dummy" subjects, together with some research assistants at CEEL (one of them played on two terminals simultaneously, in order to be able to form two groups of four players).

---

[13] This thesis is central to much recent philosophy and sociology of experimentation. See in particular Collins (1985).

1. We found a number of errors in the instructions, small typos, inconsistent terminology (e.g., sometimes we used the word *token*, sometimes *points*, sometimes *money*), unclear paragraphs, and so on. It wasn't clear, for example, what was the maximum sum that one could invest at each round.

2. Similarly, the software was affected by small typos (e.g., one window said *20* rather than *200* tokens). But it also had slightly bigger problems: sometimes the introduction to the game was too abrupt, sometimes players were left to wait for a considerable time without any explanation (wondering, for instance, whether the whole game had stalled, or was finished, or whatever else). In light of the pilot experience, we decided to add some windows with messages like "wait until all players have completed the game," or "now a new window will appear with the first round of the game," and so on.

3. When we tried to transfer all data from the server in the computer room to the server of the experimental economics laboratory, we could not do it. After a few checks and further attempts, we decided that there must have been some problem with the university server. Of course, this would have created problems had it occurred during the real experiment. One day later, we tried again and this time it worked fine.

4. Back in the lab, we had a look at pilot data. The pilot provides, among other things, a chance to check the level of payoffs. Before the experiment, we had told everyone to try to play "seriously," so as to obtain some meaningful information from the results. We found that the total payoffs were quite high compared with what we expected on the basis of previous experience and of the existing data from the literature. Our pilot subjects were more cooperative than one would expect. Of course, the sample was too small to make a reliable inference and, besides, cooperation may also have been the result of the fact that we all knew each other well and were not playing for real money. However, Roberto had had an experience of unusually high levels of contribution in public goods experiments with Italian students, and therefore, we decided to lower the payoffs to be sure of remaining within our budget limits.

5. The pilot also allows you to get a rough idea of the time length of an experiment. In our case, the pilot lasted too long. One research assistant was very careful in reading the instructions and effectively delayed the experiment by at least fifteen minutes. One may be tempted to take this event as a peculiarity of the pilot that would

hardly occur in the real experiment, but it is important to realize that the duration of each experimental session is equal to the time taken by the slowest subject in that session. With about one hundred subjects participating, it is quite likely that someone will be as slow or even slower than the research assistant in the pilot. In fact, we revised the time schedule for the experiment and increased the interval between each session as a safety measure.

6. After the pilot, we had a chat with the research assistants, none of whom had ever seen the instructions or the software before or had discussed the experiment with us. This enabled the collection of some interesting suggestions on how to improve the experiment, how to clarify the instructions, how to eliminate some useless wording, and so on. It turned out that one assistant had not understood the public goods mechanism, which induced us to add a crucial clarifying sentence in the instructions.

7. The pilot is also invaluable for data analysis. Using pilot data, we produced a few graphs and some statistics. This allowed us to check that the software we used to elaborate the raw data were reliable and could give us the information we were looking for. In technical terms, this is a procedure of "calibration": a scientific instrument (e.g., a microscope or a statistical package) is tried on some object whose properties are well known, in order to make sure that the instrument is able to give a correct representation of the object itself. In this case, the object was our behavior – a behavior that, of course, we knew very well. I knew, for example, that I had followed a certain strategy in the game and afterward, checked that this strategy emerged from the data analysis performed by the statistical tools we used.

This is just a partial list including some of the most significant findings. In reality, everything must be subjected to repeated checking: instructions, payment procedures, data collection, data analysis, and so on. Some checking must take the form of concrete pilot experiments, others can be performed by merely "simulating" the relevant procedures in the abstract (as in a thought experiment). But it is important that checks be extensive and repeated. It is quite amazing how until the day before the real experiment takes place, you keep finding small imperfections that have slipped through the net. In many cases, these probably would not cause the whole experiment to fail, but still one can never be entirely sure that it will work.

It's worth mentioning that although the phase of checking and the actual experiment have been presented as if they were sharply separated, they need not be so. Many experimenters like to run one or two sessions of the experiment for real, and then pause for a few weeks during which they make sure that the data and the design were "right." If some problem emerges from the early sessions, there is still time to remedy it. The data from the early experiments may have to be discarded, but there should remain enough time and resources to organize other sessions and complete the research anyway. Other experimenters prefer to run the experiment all at once, after extensive checking. They usually feel that adjusting one's experiment in light of the data is a form of cheating: the evidence from early sessions may prompt you to change the hypotheses under test, or to modify the design if the data do not seem to provide the result that you expected. Such procedures seem to violate some standards of scientific integrity, according to which scientists should put forward precise predictions and then test them severely. If you run the experiment all at once, the result is what it is, and no adjustments or tricks are possible. In Chapter 5, I discuss arguments concerning the importance of making "risky" predictions in more depth.

Improvising

When the checking is over, the experiment can begin. (Or perhaps, more realistically, when the experiment *has to* begin, the checking is over.) The end of the checking phase, however, does not imply that 100 percent reliability has been achieved. Something can still go wrong, and in fact, almost certainly *will* go wrong. For this reason, it is crucial to have some safety nets ready, and also enough creativity to solve problems as they come up during the experiment.

The first thing you have to do is make sure that you recruit enough subjects for the experiment. Recruiting can take many different forms, depending on the sort of population you are interested in. Most experiments are run with university students, for the simple reason that they are most readily available. The choice of such a population of course is not unproblematic, and you must be aware of the possible biases it may generate. Students are not acquainted with certain real-life problems or situations, and thus may provide you with data that are unrepresentative of the behavior you are interested in. Suppose you are studying tax evasion, for instance. A university student probably has never filled in a tax report in his or her life. For certain tasks, you might want a different population, such as a population of experienced businessmen who routinely make decisions of the sort you are investigating. On the other hand, university students may also be more trained than other people in solving certain tasks, such as highly abstract logical puzzles. You must also pay attention to *what kind* of students they are. It has been reported, to take a famous and controversial case, that students from business schools and economics departments tend to conform to standard economic theory more closely than students from other programs – which, of course, may be relevant in experiments on public goods.[14] The gender of your sample might also make a difference, and so on. The list of factors that may matter with respect to the subject population is usually large and varies from experiment to experiment.

In our case, we recruited by means of an ad placed outside the experimental lab. This obviously gave us a self-selected sample made up entirely of university students. For the sake of postexperimental analysis, we just asked our subjects to fill in a short questionnaire asking which department they belonged to, their birth date, and their gender. It is a good idea always to recruit more subjects than you need. When we had reached the numbers we were aiming for, we continued to sign on people as "replacements." If someone did not show up on the day of the experiment, we would take one of the extra students on board. This is quite important in an experiment with groups, because one student who does not show up may cause the loss of a whole group.

We planned five sessions, for a total of ninety-four subjects. The early sessions were somewhat slower. On one occasion in particular, we had to launch the software twice because for some unknown reason, it refused to start the experiment. Sometimes the server would freeze for a few seconds (up to three minutes, in one specific case), again for unknown reasons. Subjects also cause problems. Problematic subjects come in two categories: those who are embarrassed by the task and those who are not embarrassed enough. We had a couple of subjects of the first kind, who slowed the pace of the experiment considerably. One kept everybody else stuck in the room for fifteen extra minutes because he couldn't decide how much to contribute, to the utter annoyance of everybody else. When cases like these happen, you must have a plan. We had decided in advance not to intervene if possible, in order to give every subject enough time to reflect on the task at his or her own pace. In the unlikely event that we

[14] Cf. Marwell and Ames (1981) and Frank, Gilovich, and Regan (1993).

approached the end of the session (we had a fixed booking time for the computer rooms), we would start to put pressure on the slowest subjects by announcing they had only five minutes left to finish the task (and announcing then four, three, etc.). I once witnessed an experiment with a subject who was so recalcitrant she almost ruined the whole experiment. She kept saying that she could not understand the task and could not begin until she had understood it properly. Eventually, an experimenter had to sit beside her and guide her through the whole task. Of course, the experimenter wrote down the terminal code and made sure that the data from that subject and group were erased from the data set before any statistical analysis was carried out.

Subjects who are not embarrassed enough create problems of an entirely different sort. First of all, you have the super-quick ones who finish before everybody else and tend to get bored fairly soon. If they have IT skills, you are in trouble: they may try to read their e-mail while waiting for the next round, they may try to launch one of your applications in order to "have a look at the next task," and so on. In some cases, these people manage to mess up the whole experiment. We had a couple of cases like these, especially "helpful" people who began to launch applications and insert codes (the wrong ones, of course) to give us a hand. The best way to deal with such subjects is to state clearly at the beginning (and then repeat a couple of times, just to make sure) that it is absolutely forbidden to play with the terminals; it may be useful to announce at crucial times "now do not touch the keyboard until we say so," "now you can use your keyboard again," and so on.

Results

When all sessions have been completed (it took us a couple of days), the phase of data generation is over. This is not the end of the research, however. On the contrary, a most important phase of inquiry is about to begin: the extraction of results from experimental data. The information you are looking for will usually not be manifest, and part of the job of the scientist is to interrogate the data in order to obtain as much information as possible. First of all, it may be necessary to eliminate some data that you have some reason to believe may be unreliable. (A typical example is data from "problematic" subjects who have not carried out the task properly.) The remaining data are then analyzed statistically. The techniques used for this job (significance tests, correlation, regression analysis, etc.) are not the subject of this book, so I shall not get into the technical details of data analysis – which can differ considerably depending on the type of experiment.

A very common experience is that the data turn out to be richer than one would expect. An experiment is usually designed with one principal research question in mind, and the data should provide an answer to it (a clear-cut answer, hopefully). But often the evidence will contain more information – for example, in the form of new phenomena that had not been anticipated by the experimenters. In a few lucky cases, the unexpected results will be clear enough to be reported, but most often they will require a new design in order to be observed properly. The end of an experiment then becomes the beginning of a new one, which will shed further light on a related issue.

As I said, most experiments include the replication of an old result. The reason is that in order to prove that a certain factor or condition matters for a given effect, it is necessary to check the extent of the deviation from the benchmark phenomenon when that factor is added or subtracted from the standard design. In other words, experimenters typically learn by *comparing* what happens when you vary one or more conditions in a given situation or experimental design. In our case, we replicated the standard decay of overcontribution with "heterogeneous" groups before comparing it with the contribution patterns obtained in "homogenous" groups (cf. Overcontribution and Decay above). The replicated phenomenon, the decay of overcontribution, is represented in Figure 2.2 (I shall not illustrate the evidence obtained with homogeneous groups for it would divert us from our present concerns. Those interested can have a look at Burlando and Guala, in press).

If you compare the evidence obtained in our experiment with benchmark patterns like those reported in other articles or textbooks (cf. the results of Isaac et al. (1984), for example, summarized in Figure 2.1), you will notice that they are slightly different. Yet, Roberto Burlando and I felt that the decay phenomenon had been successfully replicated. One interesting question then is *how different* a phenomenon should be in order for the replication to fail. When the result to be replicated is defined in precise quantitative terms, this question can be answered using standard statistical tests. However, this is rarely the case in the social sciences, in which phenomena can often only be defined qualitatively. The definition of the decay phenomenon, for instance, leaves quite a wide room for maneuver. However, certain patterns are clearly incompatible with it: we surely would have been surprised had we observed, for instance, an

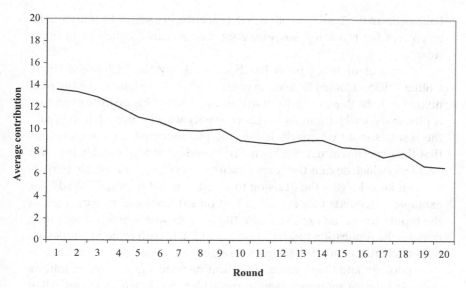

Figure 2.2. Overcontribution and decay (from Burlando and Guala, in press).

increasing level of contributions. The fact that we started with high average levels of contributions that declined smoothly throughout the experiment encouraged us to claim that the decay phenomenon had been successfully replicated.

Successful replications are relatively unproblematic, from a methodological viewpoint. The decay phenomenon is fairly easy to obtain, and has been reported so many times in the literature that no one really doubts that it can (and should) be produced in certain conditions. Failed attempts at replication raise more tricky questions: suppose we did not observe the decay phenomenon – what should have we concluded from that? The decay phenomenon is so entrenched in the literature that we surely would have questioned our capacity as experimenters in the first place; we would have probably revised some aspects of our design, and tried again until we had managed to obtain the desired effect. Some experimenters (e.g., Rubinstein 2001) argue that such a practice is strongly biased in favor of confirming other people's results, and in fact, makes the discovery of errors unlikely; but surely an initial failure would have taught us something useful and perhaps even potentially interesting from a theoretical viewpoint. Had we not observed the decay effect at the first shot, by wiggling with our design, we could have learned something about an important factor that prevented the decay from occurring; perhaps this would have been an interesting result worth reporting in a scientific journal. The incentives for reporting *interesting* failures are not negligible, in other words.

A more challenging point has been made by the sociologist Harry Collins (1985). Collins focuses on cases in which the existence of the phenomenon to be detected is itself in question (one of his favorite examples is physicists' early attempts to detect gravity waves). If we fail to obtain the result, should we conclude that the phenomenon does not exist, or that the experiment has not been performed correctly? Collins uses this problem (which he calls the "experimenter's regress") to argue that experimental knowledge – the decision to accept a result as "established," for example – depends to a crucial extent on extrarational factors, such as the reputation of the experimenter, the tenacity with which a community pursues the replication task, the funds invested in such projects, and so on. Other scientists and philosophers disagree and argue that experimental methodology, and the practice of replicating results in particular, follows rational rules of inference (see in particular the debate between Allan Franklin [1994] and Collins [1994]).

To tackle this philosophical controversy at this stage would be rather futile, because clearly the whole issue revolves around the problem of what counts as a rational scientific inference, and I have said absolutely nothing about it yet. The rest of this book is devoted to illustrating and discussing how experimenters infer from data to phenomena within an experiment, and how they generalize from such results to what happens outside the laboratory walls. Part of the job is to figure out whether these practices can be justified epistemically or, in other words, whether they can be expected to generate proper (rationally justified) scientific knowledge. I also look comparatively at the methods used in other (nonexperimental) branches of economics, and at what is done in other disciplines, such as physics and medicine, in order to spell out exactly the peculiarities of the methodology of experimental economics.

THREE

Hypothesis Testing

Like scientists, philosophers of science make use of models. Models of scientific method aim at capturing the processes of scientific reasoning, and work in many respects like scientific models: they are idealizations to begin with, and as such, they usually do not represent the real reasoning processes followed by scientists in all their details. Rather, they try to capture some "essential" aspects, and sometimes deliberately simplify in order to represent a problem or issue in a particularly forceful way. Moreover, like economic models, models of scientific method often have a hybrid descriptive-normative status, for they also aim at giving advice on the way in which scientists *ought* to reason, and help identify by way of a contrast our most common cognitive mistakes.

In this chapter, I introduce a well-known model of scientific method: the so-called Hypothetico–Deductive model of testing (or HD model, for simplicity). The HD model is an extremely useful tool to highlight some of the most fundamental problems of scientific methodology. As a representation of actual scientific reasoning, it is admittedly abstract and simplistic. Yet, it is flexible enough and can be easily amended to make room for some obvious exceptions and counterexamples. The HD model falls squarely in the empiricist tradition, according to which empirical evidence is the primary source of validation for our theories of the natural or social world. Most experimental economists would agree with this fundamental philosophical principle. Indeed, experimental economics aimed right from the start at providing better empirical tools for the discovery and testing of economic hypotheses. If scientific ideas are to be tested against the facts, the first important job is to define exactly what these "facts" are; then, I move on to describe how they can be used for testing purposes.

A taxonomy of experiments

What is a *fact* in experimental science? Facts are notoriously tricky philosophical entities, and I won't try to provide a precise definition here; at the risk of annoying some philosophers, I will speak interchangeably of "facts" and "events," pretty much as we do in everyday conversation. It is important, however, to distinguish facts from another key element of scientific method, the concept of *evidence*. A fact is a fact "in its own right," so to speak, whereas evidence is *relational*: x is or counts as evidence only in relation to one or more hypotheses. A fact, then, can be used as "evidence for (or against)" a scientific hypothesis. Because scientists test different hypotheses in different experiments, or at different stages of the same experiment, different kinds of facts count as evidence at different stages of research. What we need, then, is a classification of economic experiments, according to their purposes and motivations.

Among the taxonomies that have been proposed in the literature,[1] the most popular, and my personal favorite, is Alvin Roth's – sufficiently articulated to capture the diversity of experimental practice, but simple enough to unify many experiments under a few encompassing categories. Roth (1986, 1988, 1995) proposes a threefold classification of economic experiments based on their primary goals: *Speaking to theorists*, *Searching for facts*, and *Whispering in the ears of princes*. The first category captures all experiments aimed at testing hypotheses derived from formal theoretical models. The second, "Searching for facts," includes all experiments devoted to investigating phenomena that cannot be explained by existing theories. The third class, finally ("Whispering in the ears of princes"), includes experiments devoted to illuminating or supporting policy making. We have already seen examples of type 1 and type 2 experiments in Chapter 2: experiments on public goods began with the aim of testing the received theory, but experimenters quickly turned to investigating and trying to understand the decay of overcontribution quite independently from the existence of an alternative theoretical framework (albeit with the aim of developing one). For experiments of the third type, we shall have to wait until the second part of this book (especially Chapter 8).

---

[1] Cf. e.g., Smith (1982, 1994), Friedman and Sunder (1994), Sugden (in press).

Data and phenomena

So what sorts of facts are used as evidence in type 1 and type 2 experiments? Jim Bogen and Jim Woodward have proposed a useful distinction between *data* and *phenomena*:

Data, which play the role of evidence for the existence of phenomena, for the most part can be straightforwardly observed. However, data typically cannot be predicted or systematically explained by theory. By contrast, well-developed scientific theories do predict and explain facts about phenomena. Phenomena are detected through the use of data, but in most cases are not observable in any interesting sense of the term. (1988, pp. 305–6)[2]

Bogen and Woodward have in mind paradigmatic physical phenomena like weak neutral currents or the rate of neutrino emission from the sun, but this distinction applies to the social sciences, too. Consider our experiments on repeated public goods games. The work done by Roberto Burlando and myself belongs to a more general research program aimed at finding an adequate explanation of overcontribution and decay. In the experimental laboratory, we collected some evidence that we thought would help us in such a long-term project. Here are some *data* from our experiment:

```
EXPERIMENT_DATE:5/23/02
Session: 23/05/2002 9:50
EXPERIMENT:PG
GAME STARTED

NewRound
0:1:10:10:34:::1:2:20:0:38:::2:3:20:0:40:::3:4:3:
17:41:::4:1:20:0:24:::5:2:20:0:38:::6:3:20:
0:40:::7:4:3:17:41:::8:1:20:0:24:::9:2:15:5:43:::
10:3:20:0:40:::11:4:20:0:24:::12:1:0:20:44:::13:2:
20:0:38:::14:3:20:0:40:::15:4:20:0:24:::

NewRound
0:1:8:12:36:::1:2:20:0:40:::2:3:20:0:40:::3:4:20:
0:16:::4:1:20:0:24:::5:2:20:0:40:::6:3:20:0:40:::
7:4:1:19:35:::8:1:20:0:24:::9:2:19:1:41:::
```

---

[2] See also Woodward (1989) and Brown (1994, Ch. 7).

```
10:3:20:0:40:::11:4:0:20:36:::12:1:1:19:43:::13:2:
20:0:40:::14:3:20:0:40:::15:4:10:10:26:::

NewRound

0:1:0:20:40:::1:2:0:20:50:::2:3:20:0:40:::3:4:20:
0:30:::4:1:20:0:20:::5:2:20:0:30:::6:3:20:0:40:::
7:4:20:0:30:::8:1:20:0:20:::9:2:19:1:31:::
10:3:20:0:40:::11:4:0:20:50:::12:1:1:19:39:::13:2:
20:0:30:::14:3:20:0:40:::15:4:20:0:30:::
```

I have included in the box data from the first three rounds of a single experimental session. Each string contains the observations of one round. Three consecutive colons separate the data of different players. For each player, we collected five parameters: the player's identification number, the group she belongs to, her contribution to the public project, her contribution to the private account, and her payoff in that round. Thus, for instance, the sequence : : : 0 : 1 : 10 : 10 : 34 : : : means that player number 0 belonging to group 1 contributed ten tokens to the public account, kept the remaining ten tokens in her private account, and earned thirty-four tokens in that round. (Because sixteen players participated in that session, each string consists of $16 \times 5 = 80$ figures.)

The crucial parameter for our purposes is the contribution level (the third figure in each sequence of five). Notice that no theory we can reasonably expect to come up with is likely to be able to imply deductively that in the first round of that session we observed *exactly* the following sixteen contributions: 10, 20, 20, 3, 20, 20, 20, 3, 20, 15, 20, 20, 0, 20, 20, 20. But even if we could find such a theory, its scope would be fairly limited. As Bogen and Woodward put it, data are "idiosyncratic to particular experimental contexts, and typically cannot occur outside of these contexts" (1988, p. 317). If we were to repeat the experiment, even with the same subjects, we would almost certainly not obtain these sixteen figures in the first round. A theory explaining at such a level of detail, therefore, would be pretty useless for predictive purposes.[3] Moreover, remember that sometimes not all the data are relevant to the test of a theory. Sometimes some "weird" data are excluded because it is known that something went wrong in the experiment (one subject, say, was unable to conclude the task). Thus, not only should we not require that the theory predict exactly the data that we shall observe, we shouldn't even require it to deductively predict *all* of them.

---

[3] For a general discussion of the problem with "overfitting," cf. Forster and Sober (1994).

In fact, scientists are much less ambitious than that. Roberto and I, like all economists involved in this research program, were not concerned with explaining each single observation. We wanted to understand and explain the decay of contribution *phenomenon*. One may wonder whether we are just playing with words here. *Phenomenon* comes from the Greek *phainomenai*, "to appear," "to be manifest," "to show up." But the scientific usage of this term is not entirely consistent with its etymology: a scientific phenomenon is rarely directly observable.[4] Phenomena are rather "distilled" from observation; they are a "purified" account of what can be observed. If we were to represent visually the actual observations or data obtained in an experiment, we would have to create a massive graph with one data point for each contribution of each subject in each round of the public goods game. Of course, such a representation is not even feasible, but we could imagine a big "cloud" of tiny dots distributed unevenly across the graph.

The data are *messy*, *suggestive*, and *idiosyncratic*. They are messy because we cannot explain exactly why a given player made a given choice at a given round. To be sure, we could try to do that by disaggregating the data at the group level, or even at the individual level. But in order to account for each single data point, we would probably need to cite numerous heterogeneous factors. Some player perhaps was getting bored during round 18 and changed his strategy just because he was looking for a bit of fun. Another subject did not understand the game properly. Yet another one pressed the wrong key, and so on. On the whole, the data, nevertheless, are highly suggestive. Once organized in a graph, they indicate that some distinct pattern lies below the superficial mess. The pattern is more clearly revealed if we focus on the average contribution at each round (as we did, e.g., in Figure 2.2). The decay phenomenon now emerges from the data quite clearly.

But it is important to stress that the decay of contribution is not directly observable in the lab. What is observable are the single choices, represented as data points. The decay phenomenon has to be worked out post hoc from the observations – it is a derived entity. The decay phenomenon, to be sure, is idiosyncratic too, from some respects. If we were to repeat the experiment with a different population of subjects (e.g., with British students), we would probably obtain a slightly different rate of contribution and decay. Because "we expect phenomena to have stable, repeatable

---

[4] For a discussion of the usage of the term *phenomenon* in scientific jargon, see also Hacking (1983, Ch. 13).

characteristics which will be detectable by means of different procedures" (Bogen and Woodward 1988, p. 306), what we see in Figure 2.2 cannot be *the* decay phenomenon. It is rather an *instance* of that phenomenon, which can take different forms in different contexts. If we had to define the decay phenomenon, we would probably go for something like this: an initially fairly high rate of average contribution (roughly between forty and sixty percent of the total endowment), which slowly and smoothly converges toward (but never reduces to) equilibrium. Notice the vagueness of this description: how high should the contribution be at the start? And how low at the end? How smooth should the decay be? There is no precise answer to such questions. Certainly a decay with an abrupt trough or a peak in the middle would be considered peculiar and call for a separate explanation. But no one can specify precisely the characteristics of the "ideal-typical" decay phenomenon.

There are various reasons for this. One is that this particular phenomenon has not been satisfactorily explained yet (although every scholar has his or her own favorite story). We do not know exactly which factor or set of factors is responsible for the phenomenon, and therefore, we are unable to isolate its "essential" characteristics. The phenomenon is defined inductively, by grouping together a set of features that a class of similar experiments have in common. Here comes the second reason why the phenomenon is defined in such rough terms. As noticed in Chapter 2, the result of a public goods experiment is particularly sensitive to small variations in the experimental setting. Such sensitivity makes the precise definition of the decay phenomenon difficult, because we cannot specify exactly how differences in design reflect into differences in the level of contribution, the rate of decay, and so on (although, again, we have some rough stories about it).[5]

Description and explanation

In physics, "phenomenological" laws are distinguished from "fundamental" laws not on the basis of their observability, but on the basis of their explanatory power.[6] Similar distinctions have also been drawn by

---

[5] *Phenomena* in the sense specified here are similar to economists' "stylized facts," at least in the way in which this expression was originally used in growth theory. Since then, however, stylized facts have become a much broader notion and are often used to refer also to occasional observations or insights, qualitative generalizations, etc. In contrast, phenomena are often generated in tight experimental conditions, are highly replicable, and are more or less precisely measurable.

[6] Cf. Cartwright (1983, Ch. 6) for a philosophical discussion.

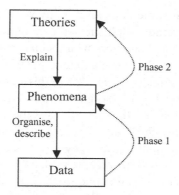

Figure 3.1. Data, phenomena, and theories.

philosophers from time to time. John Stuart Mill (1843), for example, used to distinguish between "empirical" and "causal" laws in his methodological writings. And Ernst Nagel (1961, Ch. 5), a distinguished logical positivist, uses the term *experimental laws* to denote pretty much what Bogen and Woodward call *phenomena*. Whatever terminology one is using, the key point is that phenomenological laws describe rather than explain. Phenomena organize data; data in turn do not call for a direct theoretical explanation, whereas phenomena do. Phenomena can in principle (and eventually, ideally, should) be explained by theory, whereas data result from the combination of the main causal factors modeled by theory with a number of other specific features of the data-generating process (the experimental apparatus, the instruments of observation, the materials used in the laboratory, the experimental procedures, etc.). The data–phenomena distinction can be fruitfully used to identify two separate stages in scientific research (see Figure 3.1).

The first stage ("phase 1") is aimed at organizing the observations, that is, identifying the phenomena underlying "noisy" data. This is no trivial task. Certain phenomena are so elusive and controversial that scientists do not even agree on their existence. Think of issues such as global warming, or the abnormal increase in the prices of some goods that allegedly took place in some European countries after converting their currencies into euros. Is the earth's temperature *really* increasing? Did prices in Italy go up by only 2.9 percent on average, or was the increase much more substantial, as some researchers claim? Such questions are difficult to answer because the data are often unreliable, and the very concepts that we use to interpret them are subject to debate. When an agreement is reached (if at all), it is sometimes after many years of investigation and the application of extremely sophisticated techniques of data analysis.

Phase 2 begins once a phenomenon has been identified. The job now is not only to describe (to ask *what* is happening) but also to understand or explain (*why* is it happening?). Scientists at this stage look for the *causes* of phenomena, and possibly try to organize them into theories – they seek a *systematic* explanation, in other words. The labels *phase 1* and *phase 2* do not necessarily imply chronological order. Although scientists usually try to explain a phenomenon after they have established its existence, sometimes theories are used to derive a prediction, as in the HD model – that is, to forecast that in certain conditions, a certain phenomenon will take place. After the experiment, then, the data are analyzed to check whether the phenomenon is really there. Then, if the answer is positive, the phenomenon is automatically explained by the theory from which it was derived.[7]

The role of theory

A phenomenon, however, can be "freestanding" for a long time, detached from any theory or even informal explanation of its occurrence. A successful phenomenon (i.e., a phenomenon that attracts a lot of interest and becomes the focus of research in a given discipline) is usually *surprising*, either because it goes against our commonsensical expectations, or because it contradicts a generally accepted theory. A famous example of a surprising phenomenon is the "doughnut-shaped" shadow produced by the physicist Francois Arago by projecting light on a circular disk *without* a hole in the middle. The phenomenon not only contradicted everyday experience but was also used to refute the corpuscular theory of light dominant in the early nineteenth century. Arago's disk experiment was explained using Fresnel's wave theory of light, but explanations like this are not always possible. Sometimes phenomena become somehow uninteresting to the scientific community and "die" unexplained: researchers simply stop searching for an explanation. A freestanding phenomenon (a phenomenon that cannot be presently explained), however, is always "attached" to its *design*: a description of the kind of circumstances in which it is likely to be produced and observed. In a recent paper, Robert Sugden (in press) introduces the term *exhibit* to denote the phenomenon–design pair. Borrowing an expression from Daniel Kahneman, he says

---

[7] But notice that as yet unexplained phenomena can also be used to predict; this aspect of scientific research will be more prominent in the examples discussed in later chapters.

that an exhibit is a "bottled phenomenon" – a particularly appropriate metaphor, for it reminds us that phenomena always come with their "bottle": a more or less precise description of the experimental circumstances in which it is observed.

The description of the design – the "material realization," as it is sometimes called by philosophers of experiment – is usually theory independent in the sense that it does not include or presuppose a specific explanation of the occurrence of the phenomenon. Of course, the description can and often does make use of theoretical terms (*voltage*, *potential*, etc., in physics; *risk-aversion*, *reduced lotteries*, etc., in economics), but such terms are not used to *explain* that particular phenomenon.[8] The description defines roughly the scope of the phenomenon, and hence its generalizability – but does so without making a precise commitment regarding the circumstances in which it will or will not occur (Radder 1996, 2002). Its function, rather, is to provide some aid to other researchers who intend to replicate the result in more or less similar circumstances.

The distinction between two phases in scientific research reminds us that in many cases, explanatory theories (i.e., the theories that can be used to explain a given phenomenon) do not play any significant role in phase 1. Of course, other "theories" (or hypotheses, as I prefer to call them) about the functioning of instruments, the occurrence of experimental interferences, and so on may well play a role in tests aimed at establishing phenomena. However, they often do not imply a specific explanation of the phenomenon itself. Moreover, the data–phenomena distinction can be used to highlight an important way in which scientific knowledge is cumulative. High-level, explanatory theories are often more controversial than phenomenological descriptions. Theories tend to change in time, whereas phenomena constitute a relatively more stable empirical basis to which even scientists subscribing to different theoretical paradigms can subscribe.[9]

This is particularly evident if one looks at scientific disciplines that are weak on theory but nevertheless make extensive use of the experimental method. One striking example is psychology. Psychology lacks a strong theoretical paradigm like those of physics and economics. (I am talking of

---

[8] This point is usually obscured in so-called theory-ladenness arguments in philosophy of science. For an influential discussion of the theory independence of much experimental science, cf. Hacking (1983).

[9] Since the early eighties, students of experiment have challenged the obsession with theoretical and linguistic matters that is characteristic of much recent philosophy and sociology of science. The origins of this antitheory movement can be found in Hacking (1983), but see also Cartwright (1983), Collins (1985), Galison (1987, 1997), Gooding (1990), Pickering (1995), Mayo (1996), and the papers by Bogen and Woodward cited earlier.

"theories" in a very specific sense, as structured axiomatic systems, often expressed in formal language, that unify phenomena under a coherent "umbrella" of general principles – i.e., the sort of stuff that gets published in the *Journal of Economic Theory*.) Psychologists tend to use informal hypotheses, or local models that are not unified under a common theoretical paradigm. Yet, psychology is clearly an experimental science, even more experimentally grounded than economics. Contrary to what some experimenters seem to suggest, then, the formulation of a theory is not a prerequisite for experimental science.[10]

Experimental work on public goods, to go back to my main example, was initially inspired by the desire to test the prediction of standard theory. Thus, initially, public goods experiments were aimed at theory testing. However, the majority of the work that followed does not fall in the same category. Most of the subsequent experiments were (and are) aimed at checking the relevance of small variations in the experimental conditions. Such variations range from changes in incentives and transformation functions, to changes in the subjects' sample (male and female, economics and noneconomics students, British and Italian), in the size of groups, in the quality of information, and so on.[11] Some of these conditions, to be sure, are highlighted by theory as potentially relevant to the outcome of the experiment. However, others are not: the theory is just silent about them, or in some cases, states explicitly that they are not supposed to make any difference. Yet, experimenters spend a lot of time and effort in manipulating them, just to check "what happens if . . . ," and their curiosity is often rewarded. (Recall Ledyard's analogy with table-tennis ball experiments: small differences often turn out to matter in public goods games.)

Of course, the absence of a theory to be tested does not mean that experimenters proceed randomly or blindly. Their investigations are almost invariably guided by some *hypotheses* about the behavior of their subjects. But such hypotheses often take the form of rough insights, informal guesses that are not grounded in any well-structured and rigorously formulated theoretical system and are not in the corpus of economic theory.[12]

---

[10] For two methodological accounts that link economic experiments tightly to theory, cf. Plott (1991) and Rubinstein (2001).

[11] Ledyard (1995) surveys these experiments.

[12] Hacking (1992) calls them "topical hypotheses," "to connote both the usual senses of 'current affairs' or 'local,' and also to recall the medical sense of a topical ointment as one applied to the surface of the skin, i.e., not deep" (p. 145). See also Hacking (1983, Ch. 12), in which the distinction was first introduced.

Hypothetico-deductivism

Hypothesis testing, then, is a major element of experimental research. It is fundamental both in "Searching for facts" experiments (aimed at establishing the existence of robust phenomena) and "Speaking to theorists" experiments (aimed at testing some theoretical explanation of the phenomena). The centrality of hypothesis testing is recognized in what is probably the most famous model of scientific method: the *Hypothetico-Deductive* (HD) *model of testing*. In economics, hypothetico-deductivism is usually associated with the methodological ideas of Milton Friedman and Karl Popper. Popper was a marvelous writer, and his books (especially Popper 1934 and 1963) are widely read by scientists and the general public. Friedman was partly influenced by Popper, and his popular essay "The Methodology of Positive Economics" (1953) is deeply hypothetico-deductivist in character. The HD model is rooted in a very specific empiricist tradition, which dates back to the beginning of the twentieth century.[13] According to such a tradition, science stands on two pillars: *empirical evidence* and *logic*. Evidence is the primary source of validation for our knowledge of the external world; because knowledge is expressed in linguistic statements, logic is necessary to draw inferences from one statement or set of statements to another. As we shall see, there are different views within hypothetico-deductivism about the kind of logic that is necessary and sufficient for science, from the austere attitude of those who think that standard deductive logic is enough, to the more demanding views of those who argue that we also need a logic of induction.

Let us start with the *basic HD model*. Its essential features can be summarized by means of the following two schemes:

Scheme A (refutation):	Scheme B (confirmation):
(1) $H \to e$	(1′) $H \to e$
(2) $\sim e$	(2′) e
———————	———————
(3) $\sim H$	(3′) probably (or more probably) H

In the notation I use throughout the book, the arrow " $\to$ " stands for the relation of material implication ("if . . . then . . ."), and " $\sim$ " for the

---

[13] For some recent historiography of neopositivism cf. e.g., Uebel (ed. 1991) and Michael Friedman (1999). Hypothetico-deductivism was not a feature of the early versions of neopositivist philosophy, but emerged from the work of heterodox positivists like Popper (1934) and was consolidated in the revised versions of neopositivism of the 1950s (sometimes called "logical empiricism" or "logical positivism"); see in particular Hempel (1952).

negation ("not"). The horizontal line is used to separate the premises of
an argument from its conclusions. In the two schemes above, *H* stands
for a scientific hypothesis and *e* for the empirical evidence. For instance,
to use one of the silly examples that philosophers like so much, *H* may
be the hypothesis that all swans are white, and *e* the prediction that the
two swans swimming in the lake of the local park will give birth to a little
white swan – that is, that the next swan that we shall observe in our town
will be white. In order to test the hypothesis, we must observe the color of
the newborn swan: if it is *not* white (scheme A, on the left), we conclude
that the "white swans" hypothesis is not true – in other words, *H* is *refuted*
or *falsified* by *e*. If the newborn swan turns out to be white, in contrast,
our confidence in the truth of *H* will be somehow enhanced (scheme B,
on the right).

Refutation and confirmation

Notice that there is a crucial asymmetry between the two cases: in scheme
A, the conclusion (3) follows logically from assumptions (1) and (2).
This form of argument is known as *modus tollens*, and is a valid deductive inference. In scheme B, in contrast, we cannot *by deductive logic
alone* conclude that *H* is true on the basis of (1') and (2'). Such an inference would constitute a fallacy known as "affirming the consequent." In
scheme B, in fact, I have concluded only that *H* is *probably* (or more
probably) true, given (1') and (2'). The idea, that I shall not try to make
more precise for the time being, is that the observation of *e* makes
the hypothesis *H* (from which *e* was derived) highly likely, or at least more
likely than *H* used to be before *e* was observed, or even more vaguely that
e indicates qualitatively the truth of *H*.[14] Another way of putting it is that
e confirms, or *supports*, or *indicates H*.

It is important to notice that not all fans of the HD model endorse
scheme B. Accepting B is equivalent to recognizing that the method of
science does not rely on observation and deductive logic alone. To be
able to say that a theory is confirmed by a body of evidence, you need to
make use of a logic of induction. Such logic would be *ampliative*, because
it would spell out the way in which the truth of a set of sentences *H* is

[14] These formulations point in rather different directions and are indeed captured by different formal theories of confirmation (see e.g., Giere 1977 and Achinstein 2001 for a
discussion).

indicated (or made more probable) by the truth of a set of sentences *e* implied by *H*. Or, in other words, it would specify in which way we can reach a conclusion that goes beyond the premises it is based upon.

The project of constructing a logic of inductive inference has kept busy some of the best philosophers of science of the last century. Originally, these philosophers were hoping to do for induction something similar to what had been successfully done for deductive logic at the beginning of the twentieth century: to articulate a priori the rigorous formal rules that govern this type of inference. The task is difficult, and the specialists still disagree on issues of inductive inference. Some philosophers have even cast doubt on the inductivist project as a whole. Karl Popper is probably the most famous hypothetico-deductivist rejecting scheme B altogether. According to Popper (1934, 1963), any attempt to defend an inductive logic of confirmation is doomed to fail. But this is no big deal: the logic of science for him is entirely deductive, and there is no need for confirmation or induction; *modus tollens* and refutation constitute the essence of scientific reasoning.

Popper sees the progress of science as a gigantic task of selection, in which false theories are eliminated by means of *modus tollens* and new ones are devised in order to be tested on their own. The failure to refute a theory, however, does not indicate that the theory is probable or somehow close to the truth. It just says that the theory has as yet not been proven false, according to Popper. This view has come under attack from various positions. The main problem, in a nutshell, is that scientists do not seem to propose hypotheses and theories simply in order to reject them. They also want to *use* them, to make predictions about future events or to intervene and bring about some phenomenon or state of affairs (in economics, for example, theories are used for policy purposes). As several philosophers have pointed out, in order to justify such activities, it is not sufficient to claim that a theory has not been refuted yet; at any moment in time, after all, there is, logically speaking, an infinite number of theories that have not been falsified. We would also like to think that the theory we are using is somehow better than its (infinitely many) potential rivals.[15]

Popperian attempts to dodge such an objection have reached such a level of sophistication that it would be impossible to discuss them

---

[15] Cf. e.g., Lakatos (1974) and Salmon (1988), for such a critique. Caldwell (1991) and Hausman (1992a, Ch. 10) discuss and criticize Popper's falsificationism in the context of economics in particular.

appropriately here.[16] Neither shall I discuss in depth the philosophical reasons why Popper and his followers try to defend such a stance; I shall just point out that the purely deductivist position is fairly isolated nowadays. The majority of philosophers of science believe that inductive inferences are indispensable. And the mainstream position is also closer to common sense: it seems unreasonable to hold that the fact that a theory has survived many attempts of refutation teaches *nothing at all* about its validity or future performance. However, using the past record of a theory (the fact that it was not refuted in the past) to support its use in the future (as, e.g., a predictive or policy instrument) involves an inductive or ampliative step of some sort. It is difficult to construct a good argument that turns this commonsensical intuition upside down.

The standard view of theories

Some presentations of the HD model speak of "theories" instead of "hypotheses," and of "observations" instead of "evidence." To account for Roth's distinction between theory-testing and fact-searching experiments, I prefer to avoid such terminology: the evidence may or may not be constituted by directly observable events (remember the data–phenomena distinction), and scientific hypotheses are not necessarily theoretical in character. It is true, however, that in many cases, the hypotheses belong to or are derived from theory. When this is the case, making predictions is indeed a matter of deducing testable implications from theories. Fortunately, nowadays many scientific theories are highly formalized and organized in axiomatic systems from which theorems can be derived rigorously.

I have been using the term *theory* somewhat liberally so far – but what is a theory exactly? The term is used in everyday language in many different ways. Sometimes it just means a set of abstract ideas or speculations having little to do with the concrete business of everyday life – "theories" as opposed to "hard facts." Sometimes it means something close to "hypothesis" – for example, when we say "it's just a theory" to imply that a certain claim should be taken with a pinch of salt. Such variety of meanings is also quite common in scientific language. The early proponents of the HD model, however, took the term *theory* in a very precise way. Their work

---

[16] Watkins (1984) includes a sophisticated defense of pure deductivism; Worrall (1989) provides an entertaining summary of the main arguments and a critique of the Popperian position. For a recent response to Worrall, see David Miller (2002).

on the logic of theory testing was part of a more general research program aimed at clarifying and possibly sharpening some concepts that are sometimes used liberally by scientists.[17] Once articulated more precisely, then, the concept of theory takes the form of:

(1) a set of statements postulating the fundamental entities and processes of the domain in question, and the laws that describe their behaviour;
(2) a set of "bridge principles" connecting the entities and laws to a set of phenomena to be explained and predicted.[18]

Such a "theory of scientific theories" is also known as the *standard view* or the *received view of theories* because during the 1960s, it became dominant within philosophy of science. The standard view of theories is a close relative of the HD model. The two views, to be sure, are logically independent, and it is possible to endorse one without taking the other on board. But clearly they belong to the same philosophical family and in conjunction form a fairly harmonic and coherent view of science and its method. (For this reason, by analogy, I sometimes refer to the HD model as the "standard view of testing.")

To appreciate the similarity between the HD model and the standard view of theories, consider the way in which theories explain. According to the standard view, theories are axiomatic systems and must include general laws. Such laws "cover" individual events, in the sense that they are more general than the events themselves. Thus a theory about the color of cats (if it existed) would include laws such as "The color gene is associated with the Y (male) chromosome," "The white color gene is dominant," and so on. To explain why my cat is white, for instance, I could deduce this fact from the two laws above, and the fact that my cat had a white male parent. I will say, then, that "Missy is white because she is the offspring of a white male, and the white color gene is dominant."

An explanation, in other words, would take the form of a deductive argument featuring a theory among the premises and an event or fact or phenomenon as a conclusion.[19] Given that a theory crucially includes one or more laws, and that the explanation is deductive, such a view is known as the *deductive-nomological model* of explanation (or DN model

---

[17] This sort of work is sometimes called "philosophical explication." See the classic illustration of this approach in Carnap (1950, pp. 5–6).
[18] Suppe (ed. 1977) provides a more detailed and historically sophisticated account.
[19] I am consciously simplifying here by ignoring so-called initial conditions; but see "The Duhem-Quine" problem later.

for short).[20] However, notice that the DN model has exactly the same form as the HD model. The only difference is that (i) in the HD model, the collection of the evidence is usually supposed to take place *after* the prediction has been formulated, whereas in an explanation, the evidence has usually already been observed;[21] (ii) the theoretical hypothesis is taken as a mere conjecture in the HD model (that's why we are testing it, after all), whereas in the DN model, it is taken as true or at least highly confirmed (otherwise, we could not use it to explain a phenomenon or event). The two models play different roles in the method of science – one is an account of testing, the other of explanation – but clearly go hand in hand and share a characteristic family resemblance.

The Duhem-Quine problem

Notice that so far I have hardly talked about experiments at all. This is no accident, however: the standard view (of theories, explanation, and testing) is supposed to apply to key concepts and methods of science *in general*. It assumes, in other words, that it does not matter whether one is dealing with biology or astronomy, experimental physics or archaeology: the method of testing, the nature and function of theories, and so on are supposed to be the same in all branches of science. The HD model, in particular, is aimed at characterizing the method of testing both in experimental and in nonexperimental science.

Such generality should be seen as a virtue, of course: were the HD model able to accomplish its goal, it would have done us a great service by providing a simple and unified account of scientific methodology. However, because we do draw a distinction between the experimental and the nonexperimental branches of science, we must somehow be able to account for this distinction in terms of the HD model. Suppose we want to test a hypothesis, *H*: what difference does laboratory testing make? The HD model does not say where the evidence *e* comes from. It could be collected "in the field," as well as in a laboratory experiment: the logical relations (deductive or inductive) that relate evidence to hypothesis are supposed to be the same regardless of the kind of data one is using.

[20] From the Greek *nomos*, 'law.' The classic work on the deductive-nomological model is Hempel (1965); for a recent discussion of the DN model in economics, see Mongin (2002).
[21] There is some disagreement, as we shall see, on whether a (post hoc) explanation of *e* by *H* can also count as a confirmation of *H*; but I shall keep this issue for discussion in later chapters.

The difference must lie elsewhere, then, but in order to see where, it is necessary to complicate the HD model a bit.

Scheme A and scheme B overlook a very important element of hypothesis testing. The derivation of an empirical prediction usually involves a *set* of assumptions, rather than a single hypothesis. This emerges clearly even from the simplistic examples that we have used so far. From the laws "The gene determining color in cats is associated with the Y chromosome" and "The white color gene is dominant," it is impossible to deduce that my cat is white. You also have to add to the premises the assertion "My cat has a white male parent." Statements of this sort are descriptions of empirical events or facts, and in technical jargon are called *initial conditions*. Scientific theories, therefore, can be seen as "machines" that generate empirical statements about singular events or phenomena (predictions/explanations) out of previously known empirical statements (initial conditions). Consider an astronomer trying to predict the position of a planet at a certain time t (say, tomorrow at 5 P.M.). To do that, she will have to measure the position of the planet at time t-n (say, today at 5 P.M.), and then use a set of equations or laws about the orbit of planets (Kepler's laws, or some more precise and up-to-date theory) to calculate its future position at t. The initial conditions to be measured will typically be numerous – they will include, for example, the position of other planets in the solar system, the influence of some comet, and so on. The HD model, then, should be more realistically characterized as follows:

Scheme A*	*Scheme B**
(4) $(H \,\&\, I) \rightarrow e$	(4′) $(H \,\&\, I) \rightarrow e$
(5) $\sim e$	(5′) e
(6) $\sim H$	(6′) probably H

In schemes A* and B*, the symbol I stands for a set of statements about the initial conditions that together with H are jointly sufficient for the derivation of e (a conjunction of the form $I_1 \,\&\, I_2, \ldots, \,\&\, I_n$). However, this is not the whole story yet. The application of a scientific theory requires that further assumptions be made concerning the system under study. Some theories, for instance, have "free parameters" that must be specified before they are used for predictive purposes. Economics is full of examples of this sort. Take a simple theoretical model that has been frequently used to account for overcontribution in public goods games; as in standard economic theory, agents are supposed to maximize the value of the available options, but instead of assuming narrow, selfish behavior, the value of a strategy for agent i is defined as

$$V_i = f(\pi_i, \pi_j).$$

Here, π_i is the monetary payoff of player i, and π_j is the payoff of some other player ($j \neq i$). The equation in this general form just says that agents care not only about their monetary payoffs, but also about what others achieve. In order to put it to work, it is necessary first of all to specify the form of the function relating the various factors – for example, in the following way:

$$V_i = \pi_i + a_i \sum_{j \neq i} \pi_j.$$

Of course, this equation is not specific enough yet to be empirically useful, for we also have to specify the parameter a_i, sometimes called the "altruism' factor." The parameter intuitively determines how important the payoffs of the other players are for us, by weighing their impact on V_i relative to π_i. (The "altruism" label is a bit misleading, actually, because assigning a negative value to a_i would mean that the gains of other players cause a loss of utility for i – a "spiteful" utility function, effectively.) Specifying the free parameters is a different task from measuring the initial conditions (π_i and π_j). Yet, the prediction of i's choice is dependent on the correct specification of a_i.

Another set of assumptions concerns the nonoccurrence of any phenomenon that could "disturb" the system under study and cause the prediction to fail. For instance, a macroeconomist might be able to measure the value of a parameter correctly, but the economy might be affected by some exogenous shock that causes the parameters to shift abruptly during the lag between the measurement and the event she wants to predict. Or she may neglect the influence of a factor that suddenly and unexpectedly changes the outcome. Such a problem is particularly relevant in the social sciences, in which the factors that can possibly influence a phenomenon or event are very numerous and difficult to predict. Economists tend to use simple models for reasons of tractability, and assume that no other factor except those included in the model will enter the scene. (Some factors may even be *impossible* to model because they fall outside the domain of economic theory – think of strokes for individual decision making, or political events like revolutions, wars, etc. in macroeconomics.)

It is customary to assume that the factors not modeled in the theory will either not be at work, or their influence will be small and random. A stochastic variable is normally used to capture the impact of these erratic factors; the equation above then becomes

$$V_i = \pi_i + \alpha_i \sum_{j \neq i} \pi_j + \varepsilon,$$

where α_i is the specified parameter and ε is the random ("error") variable. This procedure solves many problems but at the cost of adding more assumptions about the structure of the error term. If these assumptions are wrong, the prediction will fail.

There is yet another type of assumption that plays a crucial role in scientific testing. Experiments in the natural sciences (especially in physics and chemistry) involve the use of very complicated apparatus. Some instruments are necessary to observe entities that are too small or too remote to be detected by the naked eye. Other instruments are used to create special conditions for the experiment (e.g., bubble chambers and particle accelerators), still others to intervene on the material at hand (lasers, chemical reactors, etc.). Now, the apparatus adds complexity to the inference from the observed phenomenon to the theory. Are we sure that what we are observing is not an "artifact" of the instrument? Are we sure that the apparatus is working well? To infer that a specific hypothesis has to be blamed (or praised) for the observation of $\sim e$ (e, respectively), we need to assume that the instruments are reliable and that they have been used correctly. This may seem a problem of minor importance in economics, in which the apparatus that we use is minimal compared with, say, that of small particle physics. However, as we shall see in *experimental* economics, the role of the apparatus is nonnegligible, and many controversies revolve around the correct use and interpretation of the measurement instruments.

In order to simplify the discussion, let us call all these assumptions (about the nonoccurrence of disturbing events, the correct specification of free parameters, the structure of the error term, the correct functioning of the instruments) *auxiliary and background assumptions*. The conjunction of auxiliary assumptions (K_1 & $K_2 \dots$ & K_n) required to derive a prediction is represented by means of the letter K. The HD model then takes the form:

Scheme A**	Scheme B**
(7) $(H \,\&\, I \,\&\, K) \rightarrow e$	(7') $(H \,\&\, I \,\&\, K) \rightarrow e$
(8) $\sim e$	(8') e
(9) $\sim H$	(9') probably e

The introduction of schemes A** and B** adds realisticness to the HD model, but comes at a price. Consider A**: at a quick glance, it may look very similar to scheme A, but in fact, it differs in a crucial respect. *The argument in A** is not deductive.* Standard deductive logic does not entitle one to conclude that H is false from assumptions (7) and (8). The correct deductive conclusion must be

$$(9^*) \sim(H \ \& \ I \ \& \ K),$$

or, which is equivalent,

$$(9^{**}) \sim H \vee \sim I \vee \sim K.$$

The symbol $\vee$ stands for the logical operator *or*, and (9**) says that at least one element from the set $\{H, I, K\}$ must be false. This is as far as we can go by means of deductive logic alone. We cannot say, as I have done in A**, that the evidence $\sim e$ implies the rejection of H. The "arrow of refutation" hits the whole set of premises and cannot identify which one is responsible for the predictive failure. Was it just a mistake in the measurement of the initial conditions? Did we specify the error term correctly? Did some factor that was not included in the model mess up our prediction? Or was the main hypothesis wrong after all? These questions are important, and whatever decision we take, it had better be supported by good reasons.

The crucial role played by auxiliary assumptions in theory testing was first highlighted by the French physicist Pierre Duhem. In the fifties, the philosopher Willard Orman Quine revived Duhem's thesis in a slightly different form, and the common core of their ideas has since been labeled in the literature as the "Duhem-Quine problem."[22] The problem is not alien to experimental economists, who are in contrast well aware of its implications. Here's Vernon Smith, for example:

All tests of a theory require various auxiliary hypotheses that are necessary in order to interpret the observations as a test of the theory. These auxiliary hypotheses go under various names: initial conditions, *ceteris paribus* clauses, background information, and so on. Consequently, all tests of a theory are actually joint tests – that is, a test of the theory conditional on the auxiliary hypotheses. (Smith 1994, p. 127)

[22] Cf. Duhem (1906, Ch. 6), Quine (1953), and, on the differences between Duhem's and Quine's positions, Gillies (1993, Ch. 5). Harding (ed. 1976) includes some classical analyses of the problem. On the Duhem-Quine problem in economics in particular, see Cross (1982), Mongin (1988), and Sawyer, Beed, and Sankey (1997).

Before we start to look for a solution, it is important to specify exactly the scope of this problem, its relevance, and implications. *Whose problem* is the Duhem-Quine problem to begin with? Who is affected by it? As anticipated, there are two parties within the hypothetico-deductivist camp who disagree on the role played by inductive logic in science. Popper, as I said, is a supporter of the ultra-deductivist position, according to which science can do without a logic of confirmation. The Duhem-Quine problem is a problem especially for philosophers like Popper, because it suggests that deductive logic alone is too weak a basis to accomplish one of the principal tasks of science – theory testing.

Hypothetico-deductivists accepting both schemes A and B – those belonging to the inductivist party, that is – are less embarrassed by the Duhem-Quine problem. The presence of many assumptions regarding the premises of the HD argument presents a *challenge* (rather than a problem) for them. Such philosophers are already aware that deduction is insufficient for science, because positive inferences to a theoretical hypothesis are always *underdetermined* by the evidence.[23] For this reason, they are busy elaborating an inductive logic of testing, and it is natural for them to suppose that induction will give us a hand in solving the Duhem-Quine problem as well. We just need some rules to direct the arrow of refutation toward the right target among the premises of the deductive argument. Similarly, in scheme B**, we need to figure out whether *e* can be legitimately taken to be a success for *H*, or is a mere coincidence caused by the simultaneous failure of *H* and of other auxiliary assumptions – two errors that compensate for each other.

The Duhem-Quine problem is often presented as "the killer of the standard view," as if it were an obstacle that cannot be overcome.[24] In fact, such a conclusion is highly exaggerated; it is an insurmountable problem for a very specific version of the HD model, the ultra-deductivist one. It is true that philosophers of science still argue about the validity of specific solutions to the Duhem-Quine problem, because inductive logic is a tricky subject. The discussion of various inductive methods, in fact, occupies most of the first part of this book.

---

[23] The so-called problem of underdetermination stems from the (logical) fact that universal hypotheses have an infinite domain and therefore can never be deduced from the (always necessarily finite) empirical evidence. Because underdetermination is the flip (inductivist) side of the coin of the Duhem-Quine problem, I shall often implicitly subsume it under the latter.

[24] Cf. e.g., Hands (2001).

The role of experiment

There is a general agreement among scientists that experimentation is somehow superior to nonexperimental methods of inquiry. A crucial task for the methodologist is to turn this intuition into a more precise claim. According to the hypothetico-deductivist perspective, the difference between experimental and nonexperimental tests lies in the accuracy and reliability with which we can specify the initial and auxiliary conditions (the conjunction I & K). The laboratory allows the measurement of the initial conditions in ideal circumstances, with accurately calibrated instruments. The initial conditions, moreover, can to a large extent be set at will: if you want to observe the behavior of a certain entity or system in extreme conditions (say, in a very cold or hot environment), you can create such an environment and make sure that it is preserved throughout the experiment. You do not need to wait until nature creates these circumstances for you (which may take centuries or never happen). Similarly, for the other auxiliary assumptions: in the lab it is possible to isolate the experimental system so that no "disturbing" effects mess up your prediction, and to monitor carefully the experiment so as to spot any undesired interference. When our confidence in the initial conditions and the auxiliary assumptions is high, we say that we have achieved a high degree of "experimental control."

Such a tight inference from data can rarely be made in nonexperimental science. This problem is well known to macroeconomists and econometricians, who must deal constantly with lack of control. The critics of an econometric result often challenge the sampling procedures, blame omitted factors and hidden variables, structural changes in the economy or exogenous shocks, misspecified error terms, and so on. For this reason, controversies tend to go on for decades in macroeconomics and often fizzle out unresolved. John Hey, a prominent experimental economist, clearly alludes to this fact while discussing the problem of drawing econometric inferences from "field" data to theory:

> if "the theory" survives the test, it could be because *both* the original economic theory *and* the assumptions about the stochastic variables are correct, or because *both* the original economic theory *and* the assumptions are incorrect. There is no way of telling which. Similarly, if "the theory" does not survive the test, there is no way of telling whether this is because the economic theory is correct and the stochastic assumptions incorrect, or because the economic theory is incorrect and the stochastic assumptions correct, or because both are incorrect. Hence, a conventional econometric test of some economic theory is not really a test of that theory at all. (Hey 1991, p. 8)

Notice that from a strictly logical point of view, the same problem affects experimental tests too. Auxiliary assumptions also must be used in laboratory experimentation, which raises the issues highlighted by Duhem and Quine. However, Hey is right in suggesting that in laboratory contexts, the Duhem-Quine problem appears to be more manageable, because problems of control and theory choice can be tackled more effectively.

In later chapters, I try to illustrate how such an intuition can be formulated more precisely. For the time being, I would like to conclude this chapter with two general considerations. First, notice that the difference between experimental and nonexperimental science seems to be a matter of *degree*. In some happy circumstances, field data may allow you to draw tight conclusions, and in some cases, experimenters may face insurmountable problems of inference from data to theory. But overall, laboratory experimentation is supposed to provide more efficacious tools to tackle the Duhem-Quine problem. Secondly, one should not think that lack of control is a problem of economics or the social sciences only. There are branches of natural science in which the problem is equally daunting. Think of meteorology, for instance, or astronomy. In general, the nonexperimental branches of all sciences seem to provide less reliable results compared with the experimental ones. The problem with the social sciences lies in the variety and sheer number of factors that are at work in any single instance, and in the fact that bad theories are likely to have a very big impact upon our lives. Very few people are directly and immediately affected by a mistake in predicting the existence of a star in a very distant galaxy, whereas millions of people may suffer from miscalculating the effect of a change in the interest rate or taxation. That's why the failures of economics often appear magnified, and why it is so important to devise reliable methods for the discovery and testing of economic hypotheses.

FOUR

Causation and Experimental Control

In this chapter, I introduce a second important model of scientific method: the *perfectly controlled experimental design*. By doing so, I get at a deeper level in the analysis of experimental methodology than that which is allowed by the HD model of testing. The gain in detail will be paid for by a loss in generality, however: not all experiments are based on the perfectly controlled design. But the move is worth making for at least two reasons. First, a great number of experiments (perhaps the majority) are indirectly related to the perfectly controlled design. As we shall see, this model identifies in an abstract fashion the best possible conditions for the testing of causal hypotheses, and many experiments are indeed devoted to investigating causal relations. Second, the perfectly controlled design will turn out to be extremely useful later on (especially in Chapters 5 and 6), when I tackle issues of inductive inference and return to the Duhem-Quine problem. By looking at the sort of evidence that is provided by the perfectly controlled design, it is fairly easy to articulate more generally the requirements that a piece of evidence should satisfy in order to count as truly confirmatory for a given hypothesis. On top of that, the perfectly controlled experiment will help me introduce some basic features of causal reasoning that will turn out to be useful later in the book.

Basic principles of experimental design

Many ideas of experimental design that are routinely applied in the social sciences, psychology, and medicine derive historically from the application of statistics to agricultural trials. Following the tradition, I start with a simple agricultural example. Imagine you are a farmer and you want to test the efficacy of a new fertilizer. You could proceed as follows: take a large field and divide it into several small plots of land of equal size. Then divide the plots randomly into two groups; the plots in the first group are treated with the fertilizer, whereas the plots in the second group are not. After a few months, check whether the plots that were treated with the fertilizer are on average more productive than the plots that weren't treated. If they are, the fertilizer was efficacious; if in contrast, approximately the same amount of crop was produced in both groups, the fertilizer is probably not worth buying again.

This is a description of a simple *randomized experiment*. Despite its apparent simplicity, the example is already quite complex. To begin, let us isolate two important elements: (1) the idea that experimentation involves the comparison of two situations that differ in one respect (the application of the fertilizer), which we shall call *the treatment*; and (2) the use of *randomization* as a means to achieve uniformity before the treatment in the two situations that are to be compared. The concepts of treatment and randomization are at the core of many important scientific tests, for example, those that are routinely performed on new drugs. The methodological discussion in disciplines like medicine for various reasons tends to focus primarily on randomization. Randomization, however, seems to play a less prominent role in the methodological debates in experimental economics (fortunately, in my view). For this reason, I start by discussing point (1) and then move on to randomized trials toward the end of the chapter.

The agricultural trial sketched above is an instance of *posttest-only control group design*, a "model experimental design" represented on the left-hand side of Table 4.1. Model experimental designs feature prominently in research methods textbooks in psychology or in the social sciences. I have reproduced only three of the most common ones for discussion here, but the reader is welcome to refer to such textbooks for more examples.[1]

Some terminology is required to interpret these models. A model design consists of rows of *X*s and *O*s. Each row represents an experimental group, marked by a set of operations that are performed on it. The *O*s stand for measurements or observations of some key property or variable (say, the productivity of a crop, or the level of cooperation in a

---

[1] Cf. e.g., Dooley (2001), Christensen (2001), or Frankfort-Nachmias and Nachmias (1996). The ancestor of most textbooks of this kind is Campbell and Stanley (1963).

Table 4.1. *Three Model Experimental Designs*

Posttest-Only Control Group Design	Pretest Posttest Control Group Design	Solomon Four-Group Design
$X\ O_1$ R O_2	$O_1\ X\ O_2$ R $O_3\quad O_4$	$O_1\ X\ O_2$ $O_3\quad O_4$ R $X\ O_5$ O_6

public goods experiment), and the X marks the fact that a group (which is called the "experimental group") has been given a *treatment*. A treatment is an intervention or artificial variation imposed by the experimenter, like the fertilizer in the previous example. (In medicine, the intervention typically takes the form of a drug given to some patients; in experimental economics, it may consist, for instance, in a higher level of incentives, or in a different amount of information given to subjects.) Finally, the letter R stands for "randomized" and describes the way in which subjects are assigned either to the "experimental" or the "control" groups.

In the posttest-only control group design, the variable of interest is measured in both the experimental and control groups, only after the treatment. By comparing the two measurements of the variable of interest (or by taking the difference, $O_1 - O_2$), the experimenter can figure out whether the treatment induced some change in the experimental group. For example, if the variable of interest is the recovery rate from a disease and the treatment is a drug, by comparing recovery rates in the two groups, one can check whether taking the drug made a difference (a positive difference, one would hope) for the patients in the experimental group.

A crucial assumption behind this procedure, of course, is that the two groups were identical or at least similar enough before the experiment took place. If the control group includes patients who are more seriously ill, for instance, a posttest difference in recovery rates could misleadingly suggest that the treatment is responsible for an improvement in health conditions that is in fact to the result of totally different causes. The posttest-only model then works only under the assumptions that all the other factors or conditions that may have an influence on the variable of interest are kept "fixed" or at least are equally distributed across the two groups at the beginning of the experiment.

When this assumption is questionable or plainly not satisfied, it will be necessary to use a different design. The *pretest posttest control group design* corrects the posttest-only model by introducing a measurement *before* the treatment is administered. This has the obvious advantage of checking the level of the variable at stake. In case a difference between the two groups is detected, it is still possible to test the efficacy of the treatment by comparing the difference between the control and experimental groups *before* the intervention, with the difference *after* the intervention: $(O_1 - O_2) - (O_3 - O_4)$. Such a solution comes at a cost because a measurement is itself an intervention, and may therefore affect the result of the experiment. In order to check for the impact of pretest observations, then, one can compare the result of the experiment in a pretest posttest design with the result of a posttest-only design. The combination of these two models is known as the *Solomon four-group design*.

It would be possible to go on illustrating other, more sophisticated designs, but I prefer to stop here and reflect on the nature of these models. Each design presents in a schematic fashion a procedure aimed at solving a fairly specific experimental problem in "ideal" circumstances. Because the circumstances in which experimenters operate are seldom ideal, however, each model can be modified in order to "fix" a problem that the experimenter suspects might arise in a particular situation. A variation on the basic theme, however, often implies a trade-off in which one problem is solved at the cost of opening some other potential problems. Whether the new worry is worth taking seriously can usually be decided only on the basis of context-specific information about the experiment one is trying to make.

The perfectly controlled experiment

What is the general logic of these designs? To begin with, notice that all the models introduced above share an important feature: they are based on the logic of *comparison* and *controlled variation*. The evidence used by experimenters is usually comparative – it is evidence that a group of people (or crops, animals, entities in general) in certain experimental conditions behave differently from (or identically to) a group of people situated in other experimental circumstances. The variation in the circumstances must be carefully planned and controlled: the groups ideally should be situated in conditions that vary with respect to just *one* parameter (the treatment). This is highlighted in particular by the simplest model, the posttest-only design. The other models are proposed as second-best when the uniformity of initial conditions across the two groups is in doubt.

Table 4.2. *The Perfectly Controlled Experimental Design*

	Treatment (Putative Cause)	Putative Effect	Other Factors (K)
Experimental group	X	Y_1	Constant
Control group	–	Y_2	Constant

How can this uniformity be achieved, then? The most obvious way is by means of direct control or *matching*. Matching is the procedure of assigning to each group subjects that are identical with respect to some key characteristics, with the explicit aim of achieving groups that are as similar as possible. There are various ways in which this can be done, and again, I don't want to get into the details of the specific procedures here. As far as social science experiments are concerned, however, it is worth noticing that matching should not only take place with respect to the characteristics of the experimental subjects, but also to the conditions in which they are placed. This point is often overshadowed in standard textbook discussions because of the relative simplicity with which environmental uniformity can be achieved (by using the same laboratory and apparatus, for instance), as opposed to the difficulty of making sure that the subjects are distributed appropriately.

Whenever matching is less than perfect (and in the real world, it is always less than perfect), the differences between groups are neutralized by means of randomization. By assigning the subjects randomly to either the control or the experimental group, it is highly likely that matching mistakes or variations with respect to some unidentified factor will be distributed uniformly across the two groups (provided the sample is large enough). By using the appropriate statistical techniques, then, it is possible to check whether the differences observed post test are compatible with a pure chance effect or should be considered systematic. Randomization, in other words, belongs to the category of the "fixers," the aspects of a design that are introduced to correct for lack of direct control. Ideally, if one were able to implement the matching procedure perfectly, randomization would not be required. It is worth then specifying the ideal of the controlled experimental design in its "purest" form by amending the posttest-only design, as in Table 4.2.

Table 4.2 reproduces the essential characteristics of the posttest control group design, with some important differences: first of all, the letter R (for randomization) has disappeared. Secondly, a new column for the background factors (K_i) to be kept uniform has appeared on the right-hand side of the table. I have also slightly changed the notation and used Ys instead of Os to represent measurements. This has the advantage of bringing the notation in line with the one that is commonly used to denote independent (X) and dependent (Y) variables in mathematical and causal modeling. Table 4.2 assumes that X and Y are either present or absent, that is, that they are variables with just two possible values. But of course, it is possible to generalize and consider many-valued (even continuous) variables. The key is that we should be able to observe the difference that variations in X make with respect to Y.

Notice incidentally that the model of the perfectly controlled design does not refer to theories at all. The hypothesis "X causes Y" may well be derived from a formal theory or model, but not necessarily so. Sometimes it will be an extratheoretical or informal hypothesis about the influence of an experimental factor that is not explicitly modeled in economic theory. From this respect, the perfectly controlled experiment is totally compatible with the HD model. It simply captures the essential features of an important class of experiments, at a slightly more detailed level of description than that which can be provided by the HD framework. And it helps "unpacking" some concepts (like those of "evidence" and "hypothesis") that are left vague or generic in the HD account.

The reasoning behind the perfectly controlled design should be quite familiar: intuitively, you don't learn much by observing a static system. Imagine the electricity at home suddenly breaks down. You fear that a short circuit may be responsible for this, but where is it located? If you leave things as they are, you will not learn much. You may be lucky and spot a burnt wire, but that rarely happens. Passive observation is generally not very informative. In fact, you had better take a more active approach: switching all kitchen appliances off, for example. Now you flip the circuit breaker on, and the lights go on again; finally you turn the appliances on, one at a time. If the circuit breaker goes off when you turn, say, the toaster on, you have probably discovered the cause of the short circuit. These are homely procedures, but they form the core of experimental reasoning too.

Controlled variation is aimed at the discovery of the factors that matter for a certain experimental result or phenomenon: the factors that contribute to its instantiation, those that disrupt or break down a certain law of association, or those that make it deviate from known patterns. In order to discover or test the influence of such factors, it is necessary to compare a situation in which they are present and active with a situation in which they are not. Most often, the comparison involves a small departure from the design of a previous experiment. For example: social dilemma games are usually played under conditions of anonymity and no communication between players. If one wants to test the effect of knowing the identity of one's opponent, it will be necessary to compare the results of the game under the new circumstances, with the results obtained in the standard setting. In order to do so, one has to *replicate* the patterns of behavior that have already been established in the literature (with anonymity). Replications – already discussed in Chapter 2 – are important precisely because the method of controlled variation and the perfectly controlled experimental design require the extensive replication of known results.[2]

These considerations highlight a very important aspect of the perfectly controlled experimental design. The last column on the right-hand side reminds us that the other causal factors should be taken care of by keeping them constant across the two groups. We want to attribute the differences in Y to variations in X *only*. Of course this uniformity of circumstances may be difficult to achieve in practice. It is important to stress that the perfectly controlled experiment is an *ideal* setting. *Ideal* means that it is the best we can attain, not that it exists only in a platonic realm of ideas. In many cases, we shall be unable to realize the ideal exactly, and we shall be content with something less than perfect. When this is the case, of course, we must be aware that the inferences drawn from the experiment may be weaker than those elicited by means of the perfectly controlled design, and further checks will have to be made before reaching a definite conclusion. However, we have to live with our limitations and imperfections, in science as much as in life in general (the important thing is that we are aware and honest about them). Other books systematically examine several possible

---

[2] One may ask why it is necessary to replicate a phenomenon that is already well known. Can we not simply compare the result in the new setting with the results reported in the experimental literature? This latter procedure has the advantage of reducing the number of experimental trials and is in fact followed some of the time. But the replication of known results has an important rationale. As noticed in Chapter 2, experimental phenomena are often defined only qualitatively. Because the level of, say, cooperation in social dilemma games may vary depending on the subject population and other details of the experiment, it is important to make sure that we are comparing like with like, and this can be done only by replicating the known result. Secondly, if we discover that the new design (with the treatment) gives rise to significantly different patterns than those reported by other experimenters, we must be sure that the difference is due to the treatment only and not to some unintended difference that we have unwittingly introduced in the new design. By replicating the old result, we can check that we have not unintentionally introduced other changes in the experimental design.

less-than-ideal designs that can be used in science.[3] Here I shall limit my discussion to some of the most significant ones, like the aforementioned randomized experiment. Before I come to that, however, it is necessary to draw a link between the model of the perfectly controlled experiment and the notion of causation.

Explanation and causation

In experimental science you need variation, but not too much variation. You need variation of a special kind: factors or conditions must vary *one at a time*. The reason is simple: this special kind of variation is required in order to make sure that any correlation between dependent and independent variables (or treatments) reflects a *causal relation* between them. *The perfectly controlled experiment is the ideal design to find out about the causes of phenomena.*

This claim requires some clarification, for causation is a notion with a controversial reputation. Nevertheless, it is a concept that we use all the time, in everyday life as well as in scientific reasoning. We constantly say things like "kicking the ball caused the window panel to break," or "a rise in the interest rate caused a reduction in investments." These are instances of what philosophers call *token* causation: a particular event or set of events is identified as *the* cause of another event or set of events. Explanations of this kind seem to assume implicitly the existence of causal relations connecting types of circumstances (say, levels of the interest rate) to types of other circumstances (levels of investment). These relations are often described by means of generalizations, like "high interest rates discourage investments."

Notice that a typical characteristic of token-causal explanations is that they are usually satisfactory even though they don't provide a full account of the conditions that led to the effect. For this reason, we must be careful when we turn token-causal explanations into causal generalizations. It is not true that "kicking the ball (in such and such a way) causes the window panel to break" *in general*, or that raising the interest rate is *always* an effective way of discouraging investments.

---

[3] Campbell and Stanley (1966) and Cook and Campbell (1979) are two classic sources on less-than-ideal experimental designs in social science. For experimental economics in particular, see Friedman and Sunder (1994, Ch. 3); there is a continuum between less-than-ideal designs in the laboratory and so-called field experiments (or "natural" experiments, "social" experiments, etc. – see Harrison and List, in press).

The reason, of course, is that kicking the ball in such and such a way caused the window panel to break only because the conditions were favorable for that event to "trigger" the effect. We tend to think that a full explanation of the window breaking could in principle be constructed, provided enough information about the background conditions and the relevant laws of nature were available. The information required for a full explanation of the window breaking, for example, would include physical laws relating to the thickness of the glass, the mass of the ball, the position and trajectory of the ball, the velocity at the moment it impacts the glass, etc. From a statement of the relevant laws and the initial and background conditions, we should in principle be able to deduce exactly the *explanandum* (the event or events to be explained) as in the deductive nomological (DN) model introduced in Chapter 3. In practice, however, even using the most advanced scientific knowledge, we are rarely able to reach such a level of explanatory detail. Most scientific explanations are "sketchy" from some respects, and therefore look like the "kicking-the-ball" explanation above.

To illustrate, consider supply and demand models. The generalization "if prices increase, then demand diminishes" is either false or highly incomplete from a strict DN viewpoint. The truth of the generalization depends on the presence and stability of some other conditions included in standard supply and demand models, like the level of supply. But it also depends on a number of factors that are "in the background" of the model, such as tastes, technology, and endowments; the institutions that regulate the market; and social norms in general. The most plausible interpretation of the "law" of demand is, in other words, as a ceteris paribus law that holds only in the appropriate circumstances (other things being equal).[4]

But which conditions should count as "appropriate"? Intuitively, ceteris paribus clauses should include background causal conditions or factors that, together with the main cause, can contribute to bring the effect about or to prevent it from happening. Both scientific models and everyday explanations tend to select among the many relevant factors those that are somehow pragmatically more relevant for the case at hand. Thus, for instance, the causal relations studied by social scientists hold on a "background" of more fundamental physical and biological principles. If the biochemistry of the human brain were to change radically tomorrow,

---

[4] For a more detailed discussion of the supply and demand model from a causal perspective, see Hausman (1990).

the principles of supply and demand would probably cease to be valid. Yet, we do not cite biochemical conditions in economic models or explanations because we simply take for granted that they will continue to hold in the foreseeable future. Following J. L. Mackie (1974), it has become customary to call this set of "taken-for-granted" or "background" factors that make the relation between two factors (X and Y) possible the *causal field*.[5]

To sum up: ceteris paribus generalizations (e.g., that under the "right" conditions, an increase in demand will cause prices to go up) usually aim at capturing causal relations; the *ceteris* are other causal factors and preventatives of the effect, because causal generalizations hold relative to a certain set of background conditions.[6] The experimental method is the best tool we have to test causal generalizations, by making sure that the background conditions are appropriate for the causal relation to be reflected in the association between putative cause and effect.

Deterministic causation, screening conditions, and intervention

Causal reasoning is natural and ubiquitous. Some psychologists have even argued that humans and some other animals have strong causal intuitions built into their perceptions.[7] However, for a long time, causal notions have been treated with suspicion by philosophers and scientists alike. Causal deniers usually argue that every causal claim can in principle be replaced without loss by some other claim that makes use of less controversial notions, like functional dependence or statistical correlation.[8] After many years of philosophical discussion, there are good reasons to doubt that such "eliminativist" program can be successful. The main obstacles to a reduction of causation to functional or statistical dependence are that (a) functional and statistical dependence are symmetric relations, whereas

---

[5] Sometimes we interpret claims like "kicking the ball against the window causes the panel to break" as true generalizations because we believe (correctly or not) that the "right" conditions are *probably* instantiated. The message in such cases is that we had better not kick the ball against the window, because it is very likely to break it. However, in general, a cause brings its effect about only when the background conditions are "right." Hausman (unpublished) puts forward a plea to keep metaphysical issues of causation and pragmatic issues of causal generalizations distinct.

[6] See Hausman (1989, 1990).

[7] Michotte (1946) and Minguzzi (1961) are two "classics."

[8] Economists are probably familiar with the notion of "Granger causation," which is exactly the attempt to replace (or redefine) the commonsense notion of causation with a concept of statistical dependence or predictability. See Granger (1980), as well as Cartwright (1989, Ch. 2), and Hoover (2001, Ch. 7) for some philosophical discussion.

causation is typically asymmetric (if X causes Y, Y usually does not cause X); and (b) many variables that are statistically associated or functionally dependent are not causally related (problem of "spurious" correlations).[9] Although the consensus today is that cause and effect do not seem to be entirely *reducible* to other notions (like, say, the idea of correlated variables), it should be possible to *articulate* them (to show their relation to other, similar notions) so as to provide a more informative account of what it means for X to cause Y. This is what I try to do briefly in the rest of this chapter.

The model of the perfectly controlled experiment has a prestigious philosophical pedigree: it is an instantiation of the so-called *method of difference*, one of the four "canons of induction" proposed by John Stuart Mill in his *System of Logic* (1843). According to the method of difference,

Whatever antecedent cannot be excluded without preventing the phenomenon, is the cause, or a condition, of that phenomenon: Whatever consequent can be excluded, with no other difference in the antecedents than the absence of a particular one, is the effect of that one. (1843, Chapter VIII, section 2)

Once again, we find in this formulation the idea that we can discover the causes of a certain event or phenomenon by comparing two situations that are identical except in one particular respect. If such a difference in the initial conditions is not reflected in a difference at the level of the effect, then that factor or condition is not a relevant cause of the phenomenon at stake.

Mill says clearly that the method of difference "is more particularly a method of artificial experiment," and that it is "by the Method of Difference alone that we can ever, in the way of direct experience, arrive with certainty at causes" (1843, Ch. VII, section 3). The other three canons of induction ("by agreement," "by residues," and "by concomitant variations") are imperfect methods to be used when controlled experimentation is, for some reason, impossible or impractical. Nowadays, the procedure described by Mill is still used for causal discovery, and I use it as a starting point to reach a more precise characterization of causation. What notion of cause is presupposed by the method of difference or the model of the perfectly controlled experiment?

Let us go back to the kitchen example that I introduced earlier in the chapter, and suppose that the circuit breaker went off simply because of overcharging – the system could not supply enough current for three

---

[9] I shall not get into this debate here, but the interested reader can find excellent overviews in Hausman (1998a) and Hoover (2001).

Table 4.3. *Controlled Experiment: Deterministic Example*

	Y Short circuit	X Oven	[K_i]	
			Washing machine	Dishwasher
(a)	+	+	+	+
(b)	−	−	+	+
(c)	−	+	−	−
(d)	−	−	−	−

appliances to work at the same time (say, the washing machine, the dishwasher, and the oven). If I claim that the oven caused the circuit breaker to go off, I certainly do not mean to say that it can do it *alone*. Rather, I intend to say that *given the circumstances* (the fact that the washing machine and the dishwasher are already on), turning the oven on overcharges the circuit. Now the link with the perfectly controlled experiment should be obvious: if we compare the situations (a) and (b) depicted in Table 4.3, we can see that the factor "oven" constitutes the only difference between them, and hence is identified (correctly) as a cause of the phenomenon to be explained (the circuit breaker going off).

But is the oven a cause of the circuit breaker going off *in general*? No, not, for example, if all the other appliances are inactive (cases (c) and (d)). And conversely, we cannot *in general* work backwards from the circuit breaker going off (the effect) to the oven (the cause), because the same effect can easily be brought about by other means (e.g., by turning on simultaneously a hair dryer, a freezer, and a toaster). J. L. Mackie (1974) has summarized these features of causation nicely by means of his famous INUS account: a cause is an *Insufficient, Nonredundant* element in one or more sets of *Unnecessary* but jointly *Sufficient* conditions for a given effect to take place. An important virtue of this account is that it highlights the context dependency of causal relations (or the importance of the causal field).

The INUS framework, however, falls short of a satisfactory account of causation in one important respect: it fails to distinguish genuine from "spurious" causal associations. Consider the "Manchester hooters" counterexample originally due to Mackie himself (1974, p. 84), and represented schematically in Figure 4.1. Each day, the hooters of Manchester's factories sound (*A*) and roughly at the same time, many miles away, Londoners stop working (*B*). Nobody would say that *A* causes *B*, for the two events seem to be separate effects of a common cause (e.g., it being 5 P.M.); yet,

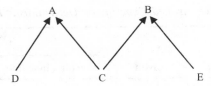

Figure 4.1. The Manchester hooters.

it can be easily demonstrated that A is an INUS condition for B.[10] This problem is typical of theories that try to reduce causation to associations between events, of which the INUS account is a sophisticated instance. A statistical analogue is the well-known paradox once proposed by David Hendry (1980): if we identify causation with statistical correlation, we end up saying absurd things like "the cumulative amount of rainfall in Scotland causes the level of prices in the UK." The mere fact that two variables happen to be highly correlated (or even deterministically associated) shouldn't induce us to claim that they are causally related.

A popular solution to such problems is to introduce a "screening-off" condition: X is a cause of Y if and only if the constant association between X and Y holds whenever the other causal factors in the background of Y are kept *fixed*. The best way to appreciate the importance of the screening-off condition is by means of a thought experiment. In the Manchester hooters example, a background factor (C = it being five o'clock) covaries with the sounding of the hooters: whenever it's not five o'clock, the hooters do not sound; whenever it's five o'clock, they do. So imagine we could achieve control of the Manchester hooters and keep them silent at five o'clock. Would the Londoners continue to work *if this were the only change introduced in the system*? Intuitively, the answer is no. C "screens off" the correlation between A and B: if we keep C "fixed" (at five o'clock) and then compare situations in which A is present with those in which A is not present (by artificially varying A), we realize that the tight association between A and B breaks down.

Notice that the screening-off condition presupposes a primitive understanding of causation, because the factors to be kept fixed in the background are defined as putative *causes* of Y. Yet, it is an informative condition, because it illustrates what sort of situations must occur in order to

[10] To see why, consider the set of *all* the conditions (D) other than it being five o'clock (C) that can make the Manchester hooters (A) sound; and the set of *all* the conditions (E) other than C that can make Londoners go home (B). Because $A \leftrightarrow C \lor D$, then $C \leftrightarrow A \& \sim D$: A is, in other words, an INUS condition for C. This together with $B \leftrightarrow C \lor E$ leads to $B \leftrightarrow \sim D \& A \lor E$: A is an INUS condition for B, QED, contrary to our intuition.

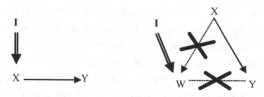

Figure 4.2. Causation and intervention.

derive reliable causal inferences from deterministic or statistical associations. Or, to put it slightly differently, it shows what we should be looking for in order to prove (or disprove) that two factors are causally connected. If we do not believe that, for example, smoking causes cancer, we should be looking for some factor in the background that explains the association between these two types of events. And once we have found such possible common cause, we should check whether keeping it fixed "in the background" is sufficient to disrupt the association between smoking and lung cancer.

Simple examples like these illustrate another important point: it is extremely difficult to disentangle our intuitions about causation and experimentation. I have already highlighted the intuitive link between causation and policy making, but clearly experimentation also has an active element in it, for the experimenter interferes with the "natural" course of events by creating the right conditions for the relation between cause and effect to become manifest in the empirical evidence. Some philosophers and statisticians have recently proposed to use a formally defined notion of *intervention* to draw the link between causation and experiments. An intervention in this technical sense is a "surgical" manipulation of a variable within a system, which leaves the rest of the system unaffected. It is possible then to define a causal relation between two factors or variables X and Y in terms of an intervention in X: X is a cause of Y if and only if it is possible to change Y by intervening in X (cf. Pearl 2000; Woodward 2002, 2003). Figure 4.2 illustrates the effect of a surgical intervention: if X is a cause of Y, the intervention (or treatment) will be reflected in variations in Y (case on the left-hand side). If, in contrast, (case on the right) two variables W and Y are spuriously correlated (e.g., because they are common effects of the same cause X), the intervention in W will not be reflected in Y, which will continue to vary depending on X. The intervention, if appropriately performed, should break the association between W and Y by interrupting the causal link between X and W.

The difference between proper experiments and other techniques of causal inference lies in the "surgical" character of the intervention. It's important to stress that the difference between "pure" experiments and other forms of scientific investigation is usually a matter of *degree*. There are several nonlaboratory circumstances in which we lack the capacity to control the relevant variables directly, and yet feel confident that a causal inference can be reliably drawn, by virtue of our "background" knowledge of the system we are studying. Consider, for example, the "natural experiment" described by Meyer, Viscusi, and Durbin (1985). In 1980 and 1982, Kentucky and Michigan, respectively, substantially raised the level of compensation for some categories of injured workers (high-earning workers). The introduction of this policy allowed a comparison not only between the recovery rates of groups of injured workers *before* and *after* the raise in benefits, but also a postintervention comparison between high-earning workers (who enjoyed the raise) and low-earning workers (who did not). Inferences from this natural experiment – time out of work increased for the workers who enjoyed higher benefits – is justified under the assumption that the policy really had a surgical effect (it didn't introduce other changes into the system), and that other independent variations in the workers' conditions (before and after the policy, say) did not occur for independent reasons.

In a "surrogate" or quasi-experimental inference, in contrast, we cannot rely even on a natural intervention such as the Michigan–Kentucky policy. As a substitute, we *select* the treatment and the control group instead of administering the treatment. Suppose you want to test the effect of studying economics on, say, cooperation in social dilemma games. Ideally, you should take a sample of people, allocate them randomly to two separate groups, and force one of the groups to study economics. However, in practice you can't do that for obvious ethical reasons. What you *can* do is separate economics students from the others and check (passively) whether they display significantly different levels of cooperation. If you observe an association between their education and level of cooperation, however, you cannot rule out that both (non)cooperative behavior and education are joint effects of a common cause (upbringing, individual attitude, genetic factors), because you have not broken the link between such factors by means of an appropriate intervention.[11]

---

[11] The example is not entirely fictional: see Marwell and Ames (1981) for the seminal observation that economics students free ride more, and Frey and Meier (2003) for an ingenious quasi-experimental attempt to identify the true causal structure behind this phenomenon.

We are now in a position to combine the ideas presented so far and propose a comprehensive articulation of the notion of causation. The key element in the "manipulationist" approach is clearly the surgical character of the intervention. An intervention in the technical sense makes the putative cause vary without changing any other relevant element in the system. But what does "relevant" mean, more precisely? It means "that it can influence the putative effect." Another way to put it is that two factors or conditions are causally related whenever they covary on a stable background of *other causal factors*:

Causation (deterministic case): *X* causes *Y* if and only if they are constantly associated in causally homogeneous background conditions.

Consistently with the model of the perfectly controlled experiment, this principle states that the association between a putative cause and its effect is a symptom of a genuinely causal relation only in some special circumstances, in which the other causal and preventative factors are not changed. According to many philosophers, this is as close as we can get to a definition of the concept of causation. To repeat: "articulations" of this sort are, of course, partly circular in character, because the notion of cause appears both on the left-hand side ("*X* causes *Y*") and on the right-hand side of the statement (the homogeneous background is defined in terms of the *causes* and preventatives of *Y*; similarly, the notion of intervention is itself clearly causal in character). However, this is to be expected: the concept of causation is primitive, and cannot be reduced to some more fundamental notion like constant association.

At any rate, I am not concerned with conceptual or ontological issues here ("what is the nature of causation?"), but with epistemic ones ("how do we know about causal relations?," "how do we infer from observable data to underlying causal relations?"). Thus, I have formulated the principle so that it can work as an inferential procedure from associations to causal relations ("if we observe such and such associations, then they reflect such and such causal relations"). Inferences of this sort require some previous knowledge about the potential factors that may distort the association between *X* and *Y*. As Nancy Cartwright (1989, Ch. 2) puts it, "no causes in, no causes out." Because our knowledge of the background factors may be incomplete and our methods of control imperfect, causal inferences from experimental data will be fallible. That's why several experiments are often required in order to reach a reliable conclusion – new experiments to improve the design of previous ones, to control for new factors, to answer new questions about some detail that had been overlooked in the past. However, epistemically there is no vicious circularity: what you need to know to find out that X causes Y does not have to do with the Xs and Ys, but with the background conditions K_i. Causal discovery is cumulative: we find out about new causal relations by relying on the ones we already know. And most importantly, the evidence collected in the laboratory supports more reliable inferences than the evidence collected in the field, because our capacity to control background factors is greater therein.

Randomization

As we have seen, experimental methodology is crucially dependent on the possibility of keeping background factors under control. When this is not possible and the "other factors" are free to vary, as in nonlaboratory circumstances, we may end up with misleading associations between putative causes and effects, that is, the evidence may not reveal the underlying causal mechanisms. But the problem is: who is going to tell us which factors we must control for and which not? How can we be sure that everything relevant has been kept constant during the experiment?

One answer immediately comes to mind: *theory* is an invaluable source of information about the factors that may be relevant in a given situation. If, for example, economic theory suggests that the quality of information and the preferences of individuals are both causally effective for a certain result – say, an exchange at equilibrium price – then we should try to control both information and preferences in the experiment. But theory, as we have already noticed, may be (and generally is) incomplete: it might omit some conditions that actually matter. As a consequence, theoretical knowledge must be supplemented by other sources. One is the informal knowledge that is part of the toolbox of every good social scientist. Even though it is not explicitly mentioned in standard economic theory, for example, experimental economists know that rational choice models may fail to predict the choices of subjects facing complicated gambles, although they accurately predict their behavior when it is regimented by an appropriate market institution. Although we do not have a precise theoretical explanation of why this is so (or what exactly distinguishes an "appropriate" economic institution), there are several informal stories that are used routinely to inform the design of experiments.

Common sense is another important source of insight about possible causal factors, and past experience in running laboratory experiments is also extremely valuable; it alerts one, for example, to those situations in which raising the level of incentives will matter or in which male subjects are likely to behave differently from female ones, and so on. Finally, careful observation helps in finding out what cannot be expected on a purely a priori basis. As we have seen in Chapter 2, pilot experiments are run precisely to correct the minor imperfections of design that might mess up the experiment. Such errors often will take the form of small details and factors that we had not anticipated, but which in fact turn out to matter with regard to the experimental effect. Sometimes the checking procedure goes on up until the stage of data analysis, when unexpected features of the data suggest that some factors might actually have influenced the result, contrary to what we had previously assumed. (Normally, another experiment with a different design has to be planned to test this hypothesis.)

I have assumed so far that tight control on each background factor is practically feasible. This assumption, however, is unrealistic. In many cases, one does not have the resources to control for all background variations, because the required design would be too complicated, too costly, incompatible with the other controls, or perhaps ethically unacceptable, or simply because one does not know the full list of the relevant background factors. In such cases (i.e., in *most* cases), experimenters rely on randomization. To recap, randomization works as follows. Suppose that an unknown gene influences the level of cooperation in public goods experiments. Not only we do not know of such gene, but we don't have the means to detect it by, say, mapping the genome of our subjects. We can, however, try to control for the influence of the gene by randomly assigning subjects to either the experimental or the control groups. If the sample is large enough, each group is very likely to include approximately an equal number of subjects with the gene and an equal number of subjects without the gene. Hence, the difference in the average level of cooperation between the two groups can be imputed to the treatment (the main putative cause) and not to this particular background factor. The nice thing is that we shall randomize not only with respect to the gene, but also with respect to other unknown factors. This is a great advantage, because as the father of randomization, R. A. Fisher, pointed out, each experimental trial differs from the others in ways that cannot even be conceived of:

it would be impossible to present an exhaustive list of such possible differences appropriate to any one kind of experiment, because the uncontrolled causes which may influence the result are always strictly innumerable. (Fisher 1956, p. 55)

As we shall see in Chapter 6, the method of randomization is not exactly equivalent to the method of matching or direct control. It is weaker, and fails in some instances where direct control succeeds. For the time being, however, let us appreciate the effects of randomization in a simple case, by looking at Table 4.3 again. Imagine that we cannot control for K directly, that is, by keeping all the background factors constant. Randomizing will have the effect of spreading cases of $+$ and $-K$ equally across the experimental group $(+X)$ and the control group $(-X)$. The randomized design will have the effect of making X (or $-X$) only *statistically* (instead of deterministically) *relevant* for the occurrence of Y ($-Y$, respectively). The occurrence of Y will be twice as likely if X occurs than if it does not.

Most real cases of experimentation are characterized by a mixture of direct control over background factors and randomization. The latter is used to control for all the "other causes" that could not be directly controlled for. For this reason, the observed regularities between the main putative cause and the effect will be almost invariably probabilistic in character.

Probabilistic causation

So far, I have considered only cases in which the cause–effect relationship is extremely tight, that is, where the (non)occurrence of the cause determines the (non)occurrence of the effect, given the right circumstances. No exceptions are allowed. Under this assumption, observed exceptions or irregularities may be a symptom of two possible cases: (1) the relationship between the two factors is not causal in character; or (2) it is causal and deterministic, but the background circumstances are not the same. However, if we relax that assumption, of course, there is another possibility, namely that (3) the background conditions are the same, but the causal relation is indeterministic. Indeterminism is usually interpreted in terms of X causing not the value of Y, but a probabilistic distribution over a range of values of Y. In other words, it is as if the cause triggered a lottery, which can give different values of Y as outcomes.

Probabilistic causation has become central in the contemporary debate on causation. The reason is twofold: first, deterministic relations can be treated as a degenerate case of indeterministic relations, the special case in which the probability of Y given X is equal to one. Second, determinism does not seem to be very common in the real world. Surely the associations between events we observe in everyday life are mostly irregular: given that the bus leaves every day from the station at eight o'clock, it is impossible to predict exactly when it will stop in front of the university building. One can only indicate a probable time of arrival, obtained by calculating an average and the spread of observations around the mean. Of course, we can speculate that if we knew each and every factor affecting the journey of the bus (the driver's style of driving, his being tired today, the fact that Mr. Jones is driving to the superstore and will slow down the traffic on Church Street, etc.) we would be able to determine exactly the time of arrival at the campus. Such knowledge would be available only to a godlike creature, of course, but this is not the point. The point is: how do we know that a set of deterministic causal relations relating to each one of these factors and the time it takes for the bus to travel from station to campus really exists?

According to some philosophers, to assume the existence of underlying deterministic relations below the surface of probabilistic phenomena is just a metaphysical prejudice.[12] Such metaphysics has been supported for centuries by the repeated successes of classical physics, which assumes deterministic relations between its variables. However, since at least the Quantum revolution, determinism does not even fit the most successful microphysics. Neoclassical economic theory, having borrowed most of its mathematical formalism from classical (prequantum) mechanics and thermodynamics, is mostly deterministic in character.[13] The stochastic variables of econometric models are therefore officially aimed at capturing the cumulative effect of unknown factors not modeled in the theory, and of errors in the measurement of the variables. But in principle, we could take the stochastic terms to model an essentially indeterministic component of economic relationships. And in fact, there have been several attempts recently to introduce probabilistic elements in economic models, for example, by representing individual preferences as intrinsically stochastic.[14]

At any rate, I am not interested in the metaphysics of causation, but in the relation between statistical evidence and causal mechanisms – an epistemological rather than an ontological issue. The deterministic

---

[12] Cf. e.g., Anscombe (1971) and Suppes (1984). Albert Einstein famously endorsed the deterministic prejudice when he claimed that "God does not play dice."

[13] See Ingrao and Israel (1987) and Mirowski (1989) for a historical discussion of the relationship between mathematical physics and contemporary economic theory.

[14] See e.g., Becker, DeGroot, and Marschak (1963) and Loomes and Sugden (1995). To be precise, there are technical differences between the "error" interpretation and the intrinsically indeterministic interpretation of stochastic terms, but I shall ignore them here. See Cartwright (1989, pp. 104–15) and Hoover (2001, Ch. 2).

analysis of the previous sections can be generalized quite naturally by means of the following principle:

Causation (probabilistic case): X causes Y if and only if $P(Y \mid X) > P(Y \mid - X)$ in some causally homogeneous background situations.[15]

Roughly, a causally homogeneous situation is one in which all the causes or preventatives of Y, other than X itself, stay fixed. Consider the stock example of the correlation between lung cancer, coffee drinking, and smoking: in the (nonhomogeneous) population of adult Britons, the correlation is there; but in a homogeneous population of smokers (or nonsmokers), it disappears. Smoking screens off the correlation between coffee drinking and cancer, but the same does not happen with the correlation between smoking and cancer, which is still there once we consider a homogeneous population of coffee drinkers (or noncoffee drinkers). The reason, intuitively, is that smoking causes cancer, whereas coffee drinking is correlated with smoking and hence is (spuriously) correlated with cancer in the nonhomogeneous population. The relevance of the distinction is obvious if we consider that in order to decrease our chance of getting cancer, we should give up smoking rather than coffee drinking.

The precise definition of the factors to be kept fixed in the background is far from trivial, but I shall ignore such problems here. The general idea is that the influence of a causal factor is reflected in the probability of its effect only in some special conditions or populations, that is, when the other background factors that may influence Y are kept constant. An important issue is whether we should impose the requirement that causes raise the probability of their effects in *all* causally homogeneous backgrounds,[16] or only in *some* of them (as in the formulation above). In order to preserve continuity with the INUS analysis and with common sense, the latter solution is preferable (striking a match does not always cause it to light – e.g., if it is damp). So, under the "right" causally homogenous background conditions, a cause raises the probability that its effect will occur and conversely, a "preventative" lowers such a probability. The link between this principle, the perfectly controlled experiment, and the method of randomization should now be obvious.

---

[15] This principle originates in Nancy Cartwright's (1983) work. See also Cartwright (1989) for some amendments, and Hausman (1998a, Ch. 9) for a general discussion.

[16] As in Cartwright's (1983) original formulation; see also Humphreys (1989). This stronger requirement is known as "contextual unanimity" and has been forcefully criticized by Dupré (1993) and others.

Conclusion

The model of the perfectly controlled design is less general than the HD model, because not all experiments are aimed at testing causal claims. Some theories are best interpreted noncausally, as providing a description of regularities or associations (standard preference theory is an example). But the diversion into causal reasoning is worth taking for a number of reasons: (1) many economic theories are to be interpreted causally; (2) several experiments are aimed at discovering the causes of as yet unexplained phenomena, or at testing their robustness to changes in the experimental conditions; (3) even the testing of noncausal theories like the standard theory of choice may involve causal reasoning, for instance, when the experiment is aimed at testing the working of instruments of observation. Finally, as we shall see, (4) the perfectly controlled design provides a *paradigmatic model of inductive inference* – an example of maximally confirmatory evidence that can be generalized to articulate a more general theory of inductive inference – a theory that is applicable to the testing of *both* causal and noncausal hypotheses.

Armed with these essential tools, in the next couple of chapters, I return to inductive inference and Duhem-Quine problems with the intent of tackling them constructively. I try to impose some requirements on the relation between a body of evidence and the hypothesis it is supposed to test, and show that these requirements capture the logic of testing actually followed by experimental economists in their everyday work. The rest of Part I is entirely devoted to these tasks.

FIVE

Prediction

A key thesis of this book is that experimental inferences can be classified in two broad categories: (1) inductions *within* the experiment and (2) inductions *from* the experiment. In Chapter 3, I introduced a further distinction borrowed from Bogen and Woodward (1988). Inductions within the experiment typically proceed in two phases: in the first phase, scientists use empirical data to identify underlying patterns or phenomena; in the second, they try to explain these phenomena by means of "deeper" theories or causal hypotheses. Both steps are nontrivial, in the sense that the data do not univocally indicate the underlying phenomena and the latter do not univocally indicate their causes. As a consequence, scientific inferences require a certain amount of ingenuity. Following an established terminology, we can say that experimental claims or hypotheses are typically *underdetermined* by the evidence. This chapter is devoted to discussing the attempt to solve this problem by means of the criterion of predictive success. I argue that such an attempt and other related projects fail because they ignore a key aspect of scientific inference: the background factors that determine whether the evidence confirms, refutes, or neither confirms nor refutes a hypothesis.

Like many other philosophers, I believe that the solution to underdetermination problems must lie in the area of inductive logic. Theories of inductive inference are numerous and at times highly sophisticated. Presently there is no agreement on a general theory, a fact that prima facie may seem to play in favor of the skeptics. However, we should not make too much of this, because fortunately theorists agree on several fundamental requirements that an adequate theory of induction should satisfy. And at any rate, my aim in this chapter is a rather modest one.

Instead of proposing a general theory of inductive inference (a daunting task, admittedly), I outline an approach to inferential problems that follows quite naturally from the practice of experimental science. The notion of the perfectly controlled experiment, in particular, plays a pivotal role in my project. Whether the proposed account is applicable outside the experimental branches of science – to econometrics or astronomy, for example – is not discussed here. For illustration, and in order to test my proposal, I use a series of examples taken from a well-known controversy within experimental economics: the case of preference reversals. To make sure the philosophical discussion accounts for the real practice of experimenters, it is necessary to dig into the details of the preference reversal experiments.

Prediction for its own sake

An unfortunate feature of human psychology is that our intuitions about inductive matters seem to be systematically disturbed by deductive biases. It is very important therefore that we resist the temptation to impose on inductive inferences requirements that are appropriate for deductive ones only. A typical example is the idea that for e to confirm H, it is necessary and sufficient that e is logically implied by H, or, in other words, that confirmation is just deduction written backward.[1] If that were the case, testing would be a totally ineffective way of discriminating among alternative hypotheses, because any piece of evidence can always be deduced from an infinite number of hypotheses, as highlighted by Duhem and (especially) Quine. The victims of deductivist intuitions tend to give up at this stage. However, the right moral to be drawn is much less dramatic: the Duhem-Quine and the related problem of underdetermination simply suggest that confirmation cannot be deduction written backward.[2] We need to impose, in other words, more stringent requirements on the relation of inductive support (the relation between e and H). But what requirements?

Many philosophers and scientists hold that e must not only be implied, but also *predicted* by the hypothesis. They impose, in other words, a

---

[1] Another way to put it is to say that scientific inferences consist of the whole body of deductive logic plus the fallacy of "affirming the consequent," but rehabilitated. "Affirming the consequent" arguments have the following form: $P \rightarrow Q$; Q; therefore P.

[2] Quine's (1953) essay is reprinted in a collection called *From a Logical Point of View*. Many commentators seem to forget the title and miss the exact implications of the Duhem-Quine problem.

temporal requirement upon the relation of inductive support: a given piece of evidence cannot indicate or confirm a hypothesis unless its observation has been forecasted (derived from *H* in advance of its observation). Take Popper, for instance: a newly proposed theory

should be *independently testable*. That is to say, apart from explaining all the *explicanda* which the theory was designed to explain, it must have new and testable consequences (preferably consequences of a *new kind*); it must lead to the prediction of phenomena which have not so far been observed. (1963, p. 241)[3]

This requirement is also implicit in the presentation of the HD model in Chapter 3, in which I highlight the symmetry between explanation and prediction: to explain is to derive from a theory/hypothesis some evidence that has already been observed; to predict is to derive some evidence that *will be* observed. The temporal requirement says that only evidence that has been successfully predicted can provide inductive support to a hypothesis.

The requirement of predictive success should sound familiar to economists, for it is strongly advocated by Milton Friedman in his famous essay on "The Methodology of Positive Economics" (1953). His arguments for the requirement, however, are not entirely convincing. Friedman holds that scientific theories are just tools to anticipate future events. Theories do not, and should not attempt to, describe the underlying structure of reality.

The ultimate goal of a positive science is the development of a "theory" or "hypothesis" that yields valid and meaningful (i.e., not truistic) predictions about phenomena not yet observed. (Friedman 1953, p. 7)

Viewed as a body of substantive hypotheses, theory is to be judged by its predictive power for the class of phenomena which it is intended to "explain." [. . .] The only relevant test of the validity of a hypothesis is comparison of its predictions with experience. (ibid., p. 8)

The view expressed in the first quotation is often called *instrumentalism*: the goal of science is to predict what happens in the natural and social world; there is no attempt to explain the "deep" mechanisms of reality.[4] That's why "explain" is between scare quotes in the second citation: because there is nothing to explain, strictly speaking. All you

---

[3] Cf. also Popper (1957).

[4] The second quote goes far beyond instrumentalism by imposing a domain restriction on the phenomena that a theory intends to account for. Many philosophical instrumentalists are concerned with any false implications of a theory and would therefore reject this restriction. I must thank Dan Hausman for pointing this out.

can do is to show that a certain theory provides good predictions, that a theory accounts for the data in this weak sense. The opposite of instrumentalism is *realism*: realists believe that science is about discovering the true structure of the (natural and social) world. Atoms, says the scientific realist, exist; or at least it is possible in principle to obtain evidence that indicates their existence. Instrumentalists, in contrast, believe they are just useful theoretical tools for prediction or whatever other goal we have in mind. They may or may not exist, but who cares? The same applies to the entities postulated by economic theory, from rational economic agents to maximizing firms, competitive markets, and so on.[5]

One problem is that the instrumentalist à la Friedman is at a loss whenever her favorite theory fails to predict well. Obviously, she will have to modify the theory in order to improve its performance, but how? Consider the simple theory: "if the barometer indicates 'rain,' then it will rain." If the barometer is well functioning, then such a theory is quite reliable for predictive purposes. However, if we begin to mess with the internal mechanisms of the barometer, or if we just leave the barometer without the appropriate service for a long time, the relation between the barometer's clock and the weather will probably break down (the barometer will cease to be a good instrument for prediction). However, to know how to keep a barometer in good condition (or even to know what it means for a barometer to be in good condition), we need to understand its internal mechanisms. We need to know, in other words, how the weather and the barometer's clock are related, what the "deeper" principles are that relate these two phenomena (cf. Hausman 1992b). The analogy to economic theories/models is obvious: we cannot make them function properly, let alone modify them effectively, unless we take a realistic attitude toward (some of) their components.

The second problem with a purely instrumentalist attitude is that scientists do not just seek predictive power generically, but predictive power of a very special kind. They look for theories and hypotheses that are robust to a particular kind of change and that can be used to derive reliable policy prescriptions. Scientists are not just interested in forecasting – in many cases, they also want to intervene. Doctors do not just want to tell you that you are going to develop certain symptoms, they are also supposed to tell you how to cure the disease that gives rise to such symptoms. Economists are not content with merely predicting an economic

---

[5] The best recent introduction and discussion of scientific realism in general is Psillos's (1999). According to some authors, the issue of realism takes a different form in economics than in the natural sciences. See e.g., Mäki (1996) and Hausman (1998b).

crisis, they also want to tell you how to prevent it. Following our barometer example, it is as if we were looking for an instrument giving advice about how to prevent storms, rather than just anticipating them. But a purely predictive tool may not be able to do that. A necessary condition to provide reliable policy advice is to know the deeper causal mechanisms that govern phenomena,[6] but then a realist attitude is more likely to fit the bill for a policy-oriented science like economics.

Prediction and confirmation

These critiques in principle do not affect the predictive success requirement for inductive inference, however, because the requirement can be justified also in noninstrumentalist ways. Among the realists who strongly support the predictive success requirement, we find, for example, Popper and his followers (e.g., Lakatos 1970, and in economics, Blaug 1980). According to Popper's methodology, it is absolutely crucial that scientists formulate risky predictions about future events. In part, this requirement is grounded on the observation of a crucial difference between "proper" science – by which Popper basically meant physics – and other "pseudoscientific" disciplines. Popper was particularly impressed by the astronomical observations by means of which the physicist Arthur Eddington refuted Newton's celestial mechanics. In 1919, Eddington organized an expedition to the island of Príncipe in West Africa, where a major eclipse was expected to take place in late May. During the eclipse, he tried to measure the deflection of light in proximity of the sun, using some distant star as a point of reference. Newtonian mechanics and Einstein's theory of relativity put forward two different predictions about the angle of deflection. Eddington's measurements seemed to support Einstein's theory and to contradict Newton's. Popper compared the "boldness" of physicists like Einstein and Eddington with the conservative attitude of pseudoscientists like Marxian theorists and Freudian psychoanalysts, who seldom put forward precise predictions about future events (and when they did, they often turned out to be wrong). Most of the evidence used to corroborate such pseudoscientific theories, Popper argued, is past evidence, already known events that are easily accommodated into the flexible framework of their theories. But such post hoc accommodations, according to

---

[6] Cf. Cartwright (1983) for the link between causal knowledge and intervention in general; for a critique of Friedman's instrumentalism along these lines see Hausman (2001).

Popper, are too easy. A hypothesis is truly confirmed by the evidence only if it takes some risk, if it faces the chance of being refuted by the evidence. Theories must "stick their neck out" and offer it to the "axe" of refutation, to use one of Popper's truculent metaphors. When a theory sticks its neck out, it is tested "severely." According to Popper, an empirical test is not *severe* if the theory implies something that is already known at the time the theory is tested. Evidence implied by the theory but not already known is said to be *independent*. Only as yet unobserved events can count as independent evidence in Popper's terms, hence the importance of successful predictions for the test of scientific theories.[7]

These ideas are popular among both experimental and nonexperimental scientists, including economists. They often take the form of a skeptical attitude toward hypotheses that have been formed post hoc, after the relevant data were collected. Here is, for example, Ariel Rubinstein, in a methodological paper devoted to theory and experiments in economics:

> Another problematic practice I would like to mention in passing is the sifting of results ex post, namely after the results are gathered. Obviously, if some pattern of behavior, from among an endless number of possibilities, is discovered in the data ex post, the results are much less informative. In the absence of rules of maintaining a research protocol one cannot check whether the results were conjectured before or after the results were obtained. (Rubinstein 2001, p. 626)

Economists who condemn post hoc theory formation usually present it as a sort of "cheating," but rarely spell out in detail what is wrong with this practice. Popper's reason to be skeptical of the theories that remain constantly "behind the facts" is that it seems too easy to do so: post hoc, a theory can be adjusted and made consistent with the evidence even if it is false ("no matter what," to use Quine's expression). If you state the theory and derive predictions from it in advance, in contrast, you are taking some *risk*. Popper, however, does not draw any connection between this sort of risk taking and inductive support. Remember that Popper was an anti-inductivist, and therefore such an omission is consistent with his overall philosophical position: you can only learn that a theory is false, not that it is (likely to be) true. But if, like most normal human beings, you believe that empirical evidence can also teach us something positive, then you

---

[7] The story of how Popper came to appreciate the importance of predictions is told in his essay "Science: Conjectures and Refutations" (Popper 1963) and in his autobiography (Popper 1976). In the text, I have followed Popper's very simplified account of the eclipse experiments, but much more realistic and interesting reconstructions can be found in Earman and Glymour (1980) and Collins and Pinch (1993, Ch. 2).

need to say something more precise about confirmation. It is necessary to link successful predictions with the (probable) truth of our hypotheses if you want to give the criterion full normative force.

There is an argument, going back to Duhem, that may be used for this purpose. In his book *The Aim and Structure of Physical Theory* (1906), Duhem compares the predictions made by a theory with the drawers of a filing cabinet. Each drawer has a different shape, and so do the phenomena (or "empirical laws," as Duhem calls them) that must be stored in the cabinet. A good theory should be able to state in advance which specific drawer will host each particular phenomenon. Suppose the cabinet/theory were manufactured without any specific plan in mind, that is, without the slightest idea of the causes behind phenomena. That new phenomena should fit in the drawers of such a cabinet/theory perfectly would strike us as a really amazing coincidence.

That, in the space left free among the drawers adjusted for other laws, the hitherto unknown law should find a drawer already made into which it may be fitted exactly would be a marvellous feat of chance. It would be folly for us to risk a bet on this sort of expectation. If, on the contrary, we recognise in the theory a natural classification, if we feel that its principles express profound and real relations among things, we shall not be surprised to see its consequences anticipating experience and stimulating the discovery of new laws; we shall bet fearlessly in its favour. (Duhem 1906, p. 28)

Duhem is a well-known antirealist, and the exact role played by this argument in his overall philosophy is controversial. However, given that this book is not devoted to exegetical matters, I'm quite happy to leave this problem to the historians. The important point is that the coincidence argument seems to draw a strong link between predictive success and the truth of scientific hypotheses. Contemporary philosophers have proposed new, more refined versions of the argument, but the fundamental intuition is still the same as in Duhem's formulation: successful prediction seems to reveal a theory's truth because the alternative hypothesis (that the theory is not even approximately true) would make the theory's predictive record a big miracle.[8]

The requirement of predictive success sounds plausible in several respects and, in fact, enjoys wide popularity among scientists and philosophers alike. It also seems to make sense of many episodes in the history of science and of the behavior of contemporary scientists. Sometimes it is

---

[8] One version of this argument is in fact known as the "no-miracles argument" for scientific realism; see Putnam (1975, p. 73), and Psillos (1999, Ch. 4) for a thorough analysis.

used as a stick to bash those "scientific" disciplines that do not display a particularly brilliant predictive record (see e.g., Rosenberg 1992, for a critique of economics heavily based on the predictive success requirement). For our purposes, it would be nice if it could be shown that the criterion accounts for the practice of experimental economists. To do this, we need to examine a representative example; I have chosen the case of preference reversals, an anomalous phenomenon that has been investigated extensively by economists and psychologists alike. To see the criterion of predictive success at work, it is necessary to engage the details of this interesting controversy.

Preference reversals

Neoclassical economic theory describes the properties of preference scales. Despite their key role right at the core of the theory, the ontological status of preferences remains quite problematic. Economists generally subscribe to the theory of revealed preferences, originally formulated by Paul Samuelson (1938). Revealed preference theory, however, can be interpreted in at least two different ways: as a theory of how our psychological dispositions (preferences) are made observable or revealed by behavior (choices), or alternatively, as an attempt to redefine or reduce preferences to observable choices. Although the latter interpretation has been prominent for some time because of the influence of behaviorism in the social sciences, it faces difficult conceptual and methodological problems.[9] Moreover, as we shall see shortly, some key debates in experimental economics appear pretty incomprehensible unless preferences are interpreted as, strictly speaking, directly unobservable. From now on, therefore, I shall use the term *preference* to denote the psychological dispositions behind individual choice.[10]

Several techniques for the indirect observation of preferences have been developed by experimenters, mostly based on the notion of willingness to pay. According to standard economic theory, the results of these procedures should all be consistent, the behavior of economic agents in all these measurement contexts being determined by their preferences. In the late sixties, some psychologists began to question the very

[9] Cf. e.g., Sen (1973, 1993) and more recently Hausman (2000); for a historico-philosophical perspective, see Mongin (2000).

[10] One's own preferences, to be sure, may be directly observable by introspection. However, this does not necessarily help when we deal, as we do in most cases, with the preferences of others.

Table 5.1. *Preference Reversal Bets (from*
Lichtenstein and Slovic 1971, p. 48)

	P-Bet	$-Bet
1	.99 Win $4.00	.33 Win $16.00
	.01 Lose $1.00	.67 Lose $2.00
2	.95 Win $2.50	.40 Win $8.50
	.05 Lose $.75	.60 Lose $1.50
3	.95 Win $3.00	.50 Win $6.50
	.05 Lose $2.00	.50 Lose $1.00
4	.90 Win $2.00	.50 Win $5.25
	.10 Lose $2.00	.50 Lose $1.50
5	.80 Win $2.00	.20 Win $9.00
	.20 Lose $1.00	.80 Lose $0.50
6	.80 Win $4.00	.10 Win $40.00
	.20 Lose $.50	.90 Lose $1.00

existence of preference scales. They conjectured that far from constituting the stable substratum from which all economic behavior arises, preferences display an unstable structure and depend heavily on the situation: they are "constructed" and vary from context to context (Slovic 1995).

Sarah Lichtenstein and Paul Slovic, two psychologists at the Oregon Research Institute, designed a two-stage test, later to become famous as the "preference reversal" experiment (PR from now on; see Table 5.1).[11] Subjects were asked in separate tasks to choose among two bets and to price them. The pairs had a common feature: they consisted of a bet with a high probability of winning a moderate amount of money but a low probability of losing a small amount (called the "P-bet"), and a bet with a low probability of winning a larger sum but a high probability of losing a smaller sum (the "$-bet"). Lichtenstein and Slovic's conjecture was that "bidding and choice involve two quite different processes that involve more than just underlying utilities of the gambles" (1971, p. 47).

In previous studies, Lichtenstein and Slovic (1968) had observed a high correlation between, on the one hand, prices and payoffs, and on the other, choices and probabilities. Their first PR experiment was conceived explicitly to produce patterns of choices such that the subjects chose the P-bet but bid more for the $-bet. As a matter of fact, such patterns were observed. The typical rate of reversals observed by Lichtenstein and

---

[11] Lichtenstein and Slovic (1971, 1973). For a nontechnical presentation of the early research on the PR phenomenon, cf. also Thaler and Tversky (1990).

Slovic, and then in later experiments, was between seventy and eighty percent. Not all reversals were of the kind expected by Lichtenstein and Slovic, though: in a standard PR experiment, 15 to 25 percent of reversals are of the nonpredicted, or "asymmetric," type (subjects choose the $-bet but are willing to pay more for the P-bet).

Psychologists tend to derive radical implications from these findings, and to deny the existence of a stable structure of preferences underlying individual behavior. In contrast, most economists would like to retain the idea of a preference scale; but taking PR seriously would make it inevitable to reconsider the properties of such a structure. The experimental evidence, in fact, challenges some crucial assumptions of rational choice theory. Denoting the strict preference relation by $\succ$ and the alternatives by $x_1, x_2, \ldots$, a preference relation is *acyclical* if it cannot be the case that $x_1 \succ x_2 \succ \ldots \succ x_n \succ x_1$ whatever the subset of alternatives $x_1, \ldots, x_n$. The preferences are *asymmetric* if it cannot be the case that $x_1 \succ x_2$ and $x_2 \succ x_1$ whatever the pair of alternatives. Asymmetry is a particular case of acyclicity, which is in turn is implied by the classic transitivity axiom of preference theory. To give up any of these basic assumptions would be very costly for standard economic theory. The principle that economic agents are rational would have to be abandoned, with obvious and disturbing normative and political consequences. However, giving up the rationality principle would also make the application of optimization techniques extremely problematic, if not impossible, and the whole body of neoclassical economics would have to be revised at its foundations.

Artifacts

The PR phenomenon is an example of a "phenomenon" in the sense specified in Chapter 3. To begin with, PRs are not directly observable. We can only observe patterns of behavior that, *once they have been interpreted*, appear prima facie incompatible with the claim that "there exists a transitive scale of preferences underlying subjects' choices." The observable data obtained in a typical PR experiment take the form of reports such as "subject *i* has chosen the P-bet when the $-bet was available and priced the $-bet higher than the P-bet." Such data are best defined as *price-choice reversals*. To obtain the PR phenomenon, one has to assume, to begin with, that pricing and choosing convey genuine information about preferences. At this stage, psychologists and economists part company. Economists presuppose that the same preference structure underlies both pricing and choosing, whereas psychologists – as already mentioned – doubt that the idea of a stable preference scale is useful at all. For reasons of simplicity, in this chapter, I mainly focus on the "economic" interpretations of the PR phenomenon and disregard the question whether preference structures exist at all. I therefore often identify the PR phenomenon with intransitive preferences rather than with the nonexistence of a preference scale (which is clearly a more general statement).[12] Following the "economic" approach, one is led to infer the existence of a genuinely intransitive preference structure. The phenomenon can then be represented as follows:

$$(PR) \quad P \succ_c \$ \succ_p P, \text{ and } \succ_c \; = \; \succ_p.$$

The inequality $\succ_c$ stands for "preference as emerging from choice," and similarly, $\succ_p$ for "preference emerging from pricing." Some theorizing has taken place on the way from the data to the phenomenon in the form above. The first inference from data to $P \succ_c \$ \succ_p P$ involves at least an assumption about the correct functioning of our "instruments of observation." The equality $\succ_c \; = \; \succ_p$ involves a commitment to the principle of *procedure invariance* – the idea that all economically relevant behavior is determined by the same preference scale and thus that all such behavior can be used as evidence for inferring the structure of preferences. Assumptions of this kind sanction the step from reports such as "subject i has chosen so and so" whereas "subject j has priced so and so" to claims about preferences, or from observed *price-choice* reversals to genuine *preference* reversals.

Notice that the term *preference* is used by both economists and psychologists when debating the results of PR experiments. Ironically, the *preference reversal* label was invented by psychologists, despite the fact that they do *not* believe in preferences in the economists' sense. Psychologists use the term in its "commonsense" meaning and originally did not distinguish between price-choice and preference reversals. Nevertheless, they did not oppose economists' shift to a more technical connotation. To economists, PRs are data seen through the filter of what we may call the "beginning" of an explanation – the low-level assumption that it makes sense to speak of stable preferences in the first place. This presupposition is still far removed from a full theoretical explanation, an explanation that in the PR case, would involve some precise claim about the *causes* of choices and pricing behavior.

---

[12] For some attempts to discriminate in the laboratory between the "economic" and the "psychological" interpretations, see Loomes, Starmer, and Sugden (1989) and Tversky, Slovic, and Kahneman (1990).

PR entered the economics literature thanks to David Grether and Charles Plott, who tried to investigate the phenomenon at the California Institute of Technology in the late seventies. Their research was driven by the suspicion that price-choice reversals may have been the product of some undetected experimental effect. In a seminal paper, Grether and Plott (1979) list thirteen possible sources of "disturbance," or thirteen possible ways of accounting for (or "explaining away") PR data. These included, among others:

1. Misspecified incentives: subjects in some of the Lichtenstein and Slovic experiments did not play for money or when they did, were not told about the value of the "points" they earned until after the experiment.
2. Income effects: if one's income increases during the experiment, one's attitude toward risk may change accordingly (e.g., a subject may go for riskier bets). Some of Lichtenstein and Slovic's experiments did not control for such an effect.
3. Indifference: if a subject is indifferent between two bets, she may choose one on one occasion and subsequently the other, and yet be perfectly rational in doing so.
4. Strategic responses: a subject may react to the expressions *selling price* or *buying price* as if she were *really* to sell or buy the bet – that is, as if she were to engage in trade and bargain with a real person. Hence, one may have the tendency to overprice a sold bet and underprice a bought bet.
5. Costs of information processing: the cost of making a reasoned choice may dominate the monetary incentives if these are not big enough. Subjects may then choose according to some "cheaper" heuristics that lead to the intransitivities.
6. Confusion and misunderstanding: the subjects may be initially puzzled by the new experimental situation, and their "irrationality" may be the result of lack of experience.
7. Experimenter's effects: psychologists are known for deceiving subjects, who therefore may engage in speculation about the "real purpose" of the experiment and try to outwit the experimenter.

Grether and Plott designed their experiment to control for these (and other) possible disturbances. The phrasing of the instructions, for example, was revised in order to avoid possible strategic responses (all words evocative of market-type behavior were eliminated), the payoffs were carefully explained in order to induce appropriate incentives, subjects were allowed to manifest indifference by checking "don't care" in the choice task, and so on. Despite great care in designing the experiment, however, Grether and Plott observed the same results produced by Lichtenstein and Slovic a few years earlier.

In many ways, the Grether and Plott experiments are representative of several other experiments carried out in the 1980s. Experiments in this early period aimed at demonstrating that PRs are an *artifact* of the experimental procedures. The concept of artifact is a widely used and yet rather vague scientific notion. The adjective *artificial* stands in opposition to *natural*, a word with clear normative flavor, and for this reason tends to have a negative connotation. An artificial object may be an imitation of an original and as such, not genuine, not "true." The artificial can be "deceitful," "insincere."[13] In science, the term is used in a number of ways, but here I am concerned with its use in relation to the data–phenomena distinction only. An interpretation of observable data is an artifact when it is not true, a mere illusion of the instruments of observation. Artifacts in this sense are a case of a potentially misleading connection between data and phenomena.

Microscopy textbooks, for example, report a long list of artifacts that the student is likely to encounter in laboratory work. Such artifacts can be divided in two categories: those that appear to be, but are not really there, and those that are real, but not originally there. Fringes caused by optical aberrations around the edges of a cell belong to the first category of artifacts. It is the microscope that generates the illusion of fringes, which, in fact, are not really there. We shall see that economists talked about the "artificiality" of the PR phenomenon in a similar way, at least in the first stage of the controversy. Bubbles on a slide, stains, scratches, and folds produced during the preparation of the assay, in contrast, belong to the second category of artifacts. They are really there but are produced by the experimental procedure. For instance, if the membranal border of an organelle seems to be interrupted somewhere, this may be a result of the chemicals used to preserve the tissue. The "natural" membrane was originally continuous, but the chemical substances used by the experimenter caused its deterioration;[14] not being aware of this fact, the experimenter might infer that it was a characteristic of the cell before any laboratory manipulation. In the second phase of the PR controversy, which we examine later in the book (in Chapter 10), the term *artifact* takes on this second meaning.

---

[13] See Hacking (1988) for an analysis of the term *artifact* along these lines.
[14] Cf. Lynch (1985, Ch. 4) for a number of examples from neurobiology.

The BDM and RLS mechanisms

The first preference reversal paper (Lichtenstein and Slovic 1971) reports three experiments. To control for possible disturbances due to lack of incentives, Lichtenstein and Slovic used in two of their designs an elicitation procedure known since the mid-sixties, the Becker-DeGroot-Marschak (BDM) mechanism. The BDM procedure can be used to elicit the selling price of any kind of commodity, and as such has been often used to observe subjects' preferences over lotteries. Preferences over lotteries are measured using the so-called *certainty equivalent*, the sum one would be willing to pay in order to play that bet. The certainty equivalent is an indirect measure, and therefore does not necessarily reflect subjects' *real* preferences. That's why a particular payoff mechanism, the BDM, is customarily used in economic experiments.

In a BDM elicitation, a subject is asked to state her reservation price, s, for a lottery (say, $[x, p; y, (1 - p)]$). The lottery is then auctioned, and if a buyer willing to bid a sum $b \geq s$ is found, the subject receives b; otherwise, the lottery is played and the subject receives a sum x or y according to the outcome. In practice, the auction is often simulated by the experimenters, who draw the bidding sum b from a uniform distribution over some relevant set. It is easy to show that a rational utility maximizer must state her real selling price – or that doing otherwise is a dominated strategy.[15]

The BDM mechanism is one of experimental economists' favorite instruments of observation. Compared with physicists, experimental economists make use of a minimal apparatus, but its function is the same. The apparatus allows the overcoming of difficult problems – for example, the collection of data about unobservable entities – but at the cost of complicating the experiment. To complicate things further, the BDM is often used in conjunction with the so-called Random Lottery Selection (RLS) procedure. In general, experimental subjects are asked to perform a number of tasks; instead of receiving an aggregate payment, each subject is rewarded according to the results of only *one* task selected at random. This procedure controls for income effects (when a subject is required to

---

[15] Let us denote with e the cash equivalent of a lottery X, such that $EU(e) = EU(X)$. If the selling price is overstated ($s > e$), then either (1) $b < e$, or (2) $b \geq s$, or (3) $e \leq b < s$. If (1) or (2) is the case, overstating the selling price (s) has no adverse consequences for the experimental subject. But if (3) is the case, she will have to play the lottery, whose expected value is e. Had she stated $s = e$, then she would have received a sum $b \geq e$ instead. Thus, overstating the selling price may lead to either equivalent or dominated payoffs for the experimental subject. The same reasoning can be applied to the symmetric case, where s is understated ($s < e$) (cf. Becker et al., 1964).

perform several tasks in a sequence, her preferences may vary because of changes in her wealth) and reduces experimental costs at the same time. If the selected task is a choice one, it is simply played out; if it is a pricing task, the BDM mechanism is used.

When Grether and Plott replicated Lichtenstein and Slovic's findings, they also used the BDM and RLS procedures. In the early eighties some economists began to argue that the PR phenomenon could have been an artifact of the BDM and RLS procedures. Charles Holt on the one hand and Edi Karni and Zvi Safra on the other independently and almost simultaneously began to investigate theoretically the robustness of these experimental procedures to violations of the axioms of expected utility theory. They pointed out that the controls used by Grether and Plott and other experimenters (e.g., Pommerehne, Schneider, and Zweifel 1982; Reilly 1982) "are appropriate if the axioms of von Neumann-Morgenstern utility theory are satisfied" (Holt 1986, p. 509). The dependence of elicitation procedures on expected utility theory was hardly a new discovery: the inventors of the BDM mechanism were aware and wrote explicitly that "the procedure is based upon the [. . .] well-known 'expected utility hypothesis'" (Becker, DeGroot, and Marschak 1964, p. 226). Karni and Safra's and Holt's innovative contribution consisted rather in showing how certain violations of expected utility *which do not concern the transitivity axiom* may nevertheless produce the *illusion* of preference reversals.

In order to understand this point, it is necessary to introduce briefly the basic elements of expected utility theory (EUT). The theory – a key building block of neoclassical economics – is a refinement of the commonsensical view that individual choices are dictated by preferences and beliefs. EUT imposes some restrictions on the structure of preferences and beliefs, in the form of *consistency requirements*. As above, we denote options or outcomes by means of the letters $x, y, z, \ldots$; probabilities by means of $p, q, \ldots$; and binary lotteries by $[px + (1 - p)\,y]$; and so on. The inequality $\succ$ stands for the strict preference relation. The axioms imposed on the preference relation are the following:[16]

---

[16] I present here a contemporary version of the axioms originally introduced by von Neumann and Morgenstern (1944), and later refined by Marschak (1950), Herstein and Milnor (1953), Luce and Raiffa (1957), and others. Alternative but equivalent axiomatizations can be given in terms of the weak preference relation ("is at least as preferred as"). The classic systematization of *subjective* expected utility theory (in which the probabilities are interpreted as subjective degrees of beliefs rather than frequencies) is Savage's (1954). For a philosophical discussion of the axioms, cf. Anand (1993).

(A1) ≻ is a weak-order relation:

$$(x \succ y) \rightarrow \sim (y \succ x) \text{ [asymmetry]}$$
$$(x \succ y \,\&\, y \succ z) \rightarrow (x \succ z) \text{ [transitivity]}$$

(A2) There exist some p and q strictly between 0 and 1 such that
$$(x \succ y \succ z) \leftrightarrow [px + (1 - p)z \succ y \succ qx + (1 - q)z]$$
[continuity]

(A3) For all p such that $0 < p \leq 1$,

$$(x \succ y) \leftrightarrow [px + (1 - p)z] \succ [py + (1 - p)z] \text{ [independence]}.$$

It is possible to prove that an individual whose preferences satisfy these principles behaves so as to maximize her own *expected utility*,

$$EU = \sum p_i U(x_i).$$

Such a proof, the "representation theorem" of expected utility, states that if an ordering relation ≻ satisfies (A1), (A2), and (A3), there exists a real utility function U (defined on outcomes) such that for every two lotteries x and y,

$$x \succ y \leftrightarrow EU(x) > EU(y),$$

where EU is defined as above. Furthermore, the class of functions that satisfy the representation condition is exactly the class of positive affine transformations of U.

The beliefs of individual agents are supposed to satisfy the consistency requirements of the standard probability calculus (the Kolmogorov axioms). According to the latter, a lottery in multiple stages can be formally reduced to a single-stage one, and expected utility theory requires that people's preferences in the multistage lottery are consistent with those in the reduced one. More precisely, the reduction principle states that subjects are indifferent between a compound lottery $A = (X_1, q_1; \ldots; X_m, q_m)$, giving a chance q_i to participate in a lottery $X_i = (x_1^i, p_1^i; \ldots; x_{ni}^i, p_{ni}^i)$, and the reduced lottery

$$R(A) = (x_1^1, q_1 p_1^1; \ldots; x_{n1}^1, q_1 p_{n1}^1; \ldots; x_1^m, q_m p_1^m; \ldots; x_{nm}^m, q_m p_{nm}^m).$$

As we have seen, in a PR experiment, the subjects are asked to perform a number of tasks; from these, one is selected at random (RLS procedure). If the selected task is a choice one, it is simply played out and the subject is paid accordingly; if it is a pricing task, the BDM mechanism is used to determine the earnings. Holt (1986) conjectures that experimental subjects see the PR experiment as a two-stage lottery. In the first stage, the task to be played out is randomly selected; in the event of pricing, there is a second stage, that is, the task is played out via the BDM mechanism. Then, Holt shows that if they apply reduction but not the independence principle, some subjects who *prefer* the $-bet to the P-bet may reverse their *choices* during the experiment. Choices observed via the RLS mechanism, then, may not reveal their true preferences.

Loosely, the principle of independence (A3) says that only the outcomes that distinguish two lotteries are relevant for the decision to be made. More precisely, it says that if x is preferred to y, then the compound lottery $(x, p; z, 1-p)$ is preferred to $(y, p; z, 1-p)$. The independence principle is considered less central than transitivity for the theory of rational choice. From a historical viewpoint, the introduction of the independence principle marked the shift from the theory of consumer choice under certainty to expected utility theory. Questioning independence, therefore – as opposed to other axioms of expected utility theory – does not have any damaging implications outside the theory of decision under risk. In fact, the independence principle was the first one to be empirically challenged in the early fifties (by the Nobel-winning economist Maurice Allais), and later research confirmed that individuals tend to violate the axiom regularly in certain experimental circumstances.[17] Moreover, the mathematical structure of rational choice theory dictates an implicit hierarchy among its axioms: the principle of independence becomes somehow redundant (or "inefficiently precise") when imposed on a non–well-ordered relation. In this sense, one may say that independence "implies" or presupposes transitivity (Mongin 1988). Finally, the independence principle is normatively weaker than transitivity, which can be justified by means of more direct arguments (cf. Guala 2000b).

More or less at the time when the Holt and the Karni and Safra arguments were being developed, other economists were busy constructing alternative models that could account for the known violations of expected utility theory. Mark Machina's Generalized Expected Utility Analysis (GEUA), for example, relaxed independence and allowed

---

[17] In the classic experiment known as the "Allais paradox," subjects are asked to choose first between (A) one million for sure and the lottery and (B) five million with probability 0.10, one million with probability 0.89, or 0 with probability 0.01; then they are asked to choose between (C) five million with probability 0.10 or 0 with probability 0.90 and (D) one million with probability 0.11 or 0 with probability 0.89. Many subjects choose A and C, thus violating the independence principle of expected utility theory. Cf. Allais (1953); the main violations of independence are reviewed in Camerer (1995).

utility functions to be merely differentiable rather than strictly linear in the probabilities. Other approaches, such as Chew and MacCrimmon's Alpha Utility theory and Quiggin's and Yaari's Expected Utility theory with Rank-Dependent Probabilities (EURDP), were being developed that similarly made do without independence or related principles.[18] It was therefore somehow natural (consistent with the "spirit of the time," so to speak) to argue that the illusion of PR resulted from violations of the Allais kind. According to Karni and Safra (1987), the BDM mechanism may be perceived by subjects as a two-stage lottery giving, among its outcomes, the possibility of playing out the priced gamble. Karni and Safra argue that if the independence principle is *not* obeyed, then it is not true that *always* setting the selling price equal to the certainty equivalent of the lottery maximizes its value (for the interested reader, I provide a more detailed description of the argument in Appendix A).

A number of generalized theories without independence (Karni and Safra call them "Ω-theories"; I have mentioned some of them earlier) can in principle be used to explain reversals. Karni and Safra constructed an example of how preference reversals are implied by a version of Quiggin's and Yaari's generalized EURDP, given a particular set of lotteries and initial conditions. The very pattern of choices observed by Lichtenstein and Slovic, Grether and Plott, and others can be accounted for by applying EURDP to the BDM elicitation. If agents were EURDP maximizers, the data produced by means of the BDM mechanism would not be inconsistent with the transitivity of the underlying preferences, and the PRs turn out to be "illusory."[19]

One can find other, similar arguments in the theoretical literature of the same period. By focusing on the *reduction principle*, Uzi Segal (1988), for example, argues that violations of independence may not be the only causes of the PR "illusion." His argument, again, moves from the assumption that the experimental subjects perceive their task as a two-stage lottery, and proceeds to show that *if* their beliefs do not satisfy the *reduction* principle, when they deal with particular pairs of bets, again, they may price items in a way that would not reveal their true preferences.

Back to Duhem

The critiques of Holt, Karni and Safra, and Segal raise a typical Duhemian problem. According to Duhem, a scientific prediction can be made only

[18] See Chew and MacCrimmon (1979), Machina (1982), Quiggin (1982), Yaari (1987).
[19] Karni and Safra show also that a large class of BDM-like devices would be useless for eliciting nonlinear preference relations.

by putting to work a "whole theoretical scaffolding" (1906, p. 185). It is customary nowadays to interpret Duhem's thesis broadly and include among the premises a number of different background factors: assumptions about the functioning of the instruments, the noninterference of disturbing factors (or fulfilment of the ceteris paribus clause), the correct specification of the initial conditions, and so on. When we seem to have produced a phenomenon contradicting our predictions, we cannot *by deductive logic alone*, argue for the falsification of any particular one of the assumptions involved (although we know that at least one must be false).

In our case, a number of inferences lead from the data collected in the laboratory to the PR phenomenon. Some of these inferences rely on the correct functioning of the instruments used in the experiment. If the latter are challenged, the inferential device breaks down. What are we allowed to infer from the data? Is not the very existence of the PR phenomenon questioned? The challenge in this case is made powerful by at least two factors: first, Duhem's problem is not stated in the abstract but concretely by specifying the way in which the inference from the data to the phenomenon may be mistaken. Second, the critiques point to a problem of *circularity*. The "instruments of observation" (elicitation) used in the experiments on individual choice rely upon those theories of behavior in whose investigation they are involved. Mechanisms such as the BDM procedure work by constructing further problems of choice under risk of the same kind as those under test. The phenomenon at stake is inconsistent with EUT, but the instruments used to observe the phenomenon are constructed on the hypothesis that EUT is correct.

Predictive success and preference reversals

Does the predictive success criterion fit the case of PR? Given its popularity, in a way it would be extraordinary if the criterion could not capture at least some important aspects of real scientific practice. Thus, before we examine its shortcomings, let's have a look at those parts of the PR case in which the criterion seems to work well. The situation, in a nutshell, is the following: in the early eighties, PRs constituted the main challenge to the transitivity axiom of choice theory, but an alternative account of the phenomenon was devised that questioned the instruments of observation (the BDM and RLS procedures) customarily used to elicit preferences in the lab. Such an account, however, according to the predictive success criterion, can receive confirmation only from as yet unobserved facts. In fact, Karni and Safra's paper did not just indicate that price-choice reversals do not logically imply the violation of transitivity. It also included an alternative theoretical model to account for PR data. The model is a specific version of EURDP. According to the predictive success criterion, economists should now derive new predictions from the model in order to test its validity as an alternative account of the PR data.

It is not clear whether Karni and Safra subscribe to the predictive success criterion, for here and there they seem to argue that their proposed model already receives confirmation by "old" PR evidence. Although the general tone of the paper suggests that their model is mostly illustrative, in certain passages they seem to take it very seriously, as if they had provided reasons to believe that subjects do violate independence and the PR phenomenon is indeed an artifact of the BDM mechanism. They write, for example: "what Grether and Plott tried and – as our discussion indicates, *failed to do* – is to observe, by means of [the BDM method], the certainty equivalents of given lotteries" (1987, p. 676, my emphasis). In a footnote, Karni and Safra also compare their contribution to Holt's: the latter pointed independently to violations of intransitivity, but "however, did not present an alternative theory *explicating* the 'PR' phenomenon" (p. 676, n. 4, my emphasis).

Perhaps Karni and Safra were influenced by the general consensus achieved by Quiggin and Yaari's EURDP among decision theorists. What they actually thought, at any rate, is not so important after all. What really matters is whether in principle their post hoc explanation of the data could receive support from the old PR data. According to the predictive success criterion, it should not, and it is easy to see why. Generalized expected utility theories do not claim that individuals always violate independence – they just say that they might. Such theories display in their general form several free parameters that have to be fixed in order to derive precise implications about subjects' behavior. For some values of these parameters, the consequences of EURDP are identical to those of expected utility theory. In other words, EURDP may well be true and yet the subjects do not violate independence when choosing among the lotteries used in PR experiments. Some *specific models* must be employed to account for PR – models such as the one devised by Karni and Safra.

The models are obtained by ad hoc specification: the Karni and Safra reinterpretation of the BDM procedure holds only for some particular pairs of lotteries and some values of the free parameters of the basic EURDP theory. According to EURDP, the value V of a lottery

$(x_1, p_1; \ldots; x_n, p_n)$ is given by

$$V(x_1, p_1; \ldots; x_n, p_n) = \sum_{i=1}^{n} u(x_i) \left[f\left(\sum_{j=i}^{n} p_j \right) - f\left(\sum_{j=i+1}^{n} p_j \right) \right].$$

The parameter u is the traditional monotonic increasing real-valued function defined on some interval in the real line (i.e., on a range of monetary prizes). Compared with expected utility theory, EURDP has got one extra free parameter, namely the "probability transformation function" f. Karni and Safra (1987) show that *if* the following specifications are chosen for f and u,

$$f(p) = \begin{cases} 1.1564p, & 0 \le p \le 0.1833 \\ 0.9p + 0.047, & 0.1833 \le p \le 0.7 \\ 0.5p + 0.327 & 0.7 \le p \le 0.98 \\ p, & 0.98 \le p \le 1, \end{cases}$$

$$u(x) = \begin{cases} 30x + 30, & x \le -1 \\ 10x + 10, & -1 \le x \le 12, \\ 6.75x + 49, & 12 \le x, \end{cases}$$

then for lotteries like those used by Grether and Plott – that is, $(-1, 1/36; 4, 35/36)$ and $(-1.5, 25/36; 16, 11/36)$ – the choice-price reversals can be accounted for. (Remember: the choice-price reversals, or "announced price reversals" as Karni and Safra call them, are the *data* to be explained as opposed to the allegedly artifactual PR *phenomenon*.) Notice that the particular parameters and initial conditions are doing a lot of work here. EURDP cannot by itself even account for the particular asymmetries of observed reversals: only the model with its specific parameters can. But given that the parameters are specified in the light of PR data, it is hardly surprising that the specified model can be used to account for the latter. Because Karni and Safra's model (EURDP plus specification of the free parameters plus initial conditions) was devised explicitly to account for the evidence to be explained, in other words, the latter cannot provide much support to the violation of independence hypothesis.

So the Karni-Safra argument discredits the Lichtenstein-Slovic interpretation of PR experiments only indirectly, by proposing a possible alternative interpretation of the evidence. In a way, we already knew that some such interpretation could be constructed, at least in principle. Once the alternative takes a concrete, specific form, as in Karni and Safra's paper, we need to use the empirical evidence to test it against the standard interpretation. And the old PR data do not seem to be able to do that.

Where have the auxiliaries gone?

Thus some aspects of the PR case can be described using the predictive success requirement. But one thing is to describe what happens, another is to capture the deep motivation behind a piece of scientific reasoning. To put it slightly differently: is the Karni-Safra model's lack of support to be explained by the fact that it simply accommodated old evidence, or was this just a by-product of some deeper problem? In order to answer this question, we have to examine the predictive success requirement from a normative viewpoint (rather than a purely descriptive one, as I have done so far), and ask whether it is a necessary and sufficient condition for inductive support. I will discuss three problems with this requirement, two of which are particularly related to experimental testing and one to inductive support in general. I shall start from the latter, an argument originally due to Deborah Mayo.[20]

Mayo (1996, Ch. 8) examines some counterexamples of the predictive success criterion – some instances, that is, in which a hypothesis seems intuitively to be confirmed by the evidence despite the fact that the latter was clearly used in the process of hypothesis construction. First, she asks us to consider detective investigations. In most cases, the fundamental hypothesis – that the culprit is or is not x – is constructed using (indeed is tailored upon) the available evidence. Yet, this is not usually taken by the jury as an argument *against* the hypothesis itself – quite the contrary. (A detective who *first* formulates the hypothesis that Ann killed Bob and *then* looks at the evidence would surely be accused of being prejudiced against Ann.) This argument can be extended by looking at those scientific disciplines that willy-nilly have to rely almost entirely on "past" evidence. Take archaeology or paleontology: the fact that a theory or hypothesis is modeled on the available fossils is taken to be a virtue not a defect in these fields. Here's another simple counterexample proposed by Mayo (1996, pp. 271–3): consider the hypothesis that "the average SAT score in my

---

[20] Mayo's (1996) arguments are strictly speaking directed against a modified version of the predictive success criterion – the "independent evidence" or "construct–independence" criterion, as it is sometimes called – proposed by Elie Zahar (1976, 1983), John Worrall (1978, 1985), and Ronald Giere (1983) as a response to some other, more basic objections to predictive success. Since the arguments refute both the general (predictive success) and the special version (construct–independence) of the requirement, I shall ignore such details here. There has been some discussion on the construct–independence criterion in the economic methodology literature (cf. e.g., Hands 1985), which however focuses primarily on descriptive matters and in my view misses some of the crucial normative issues (see also Salanti 1994).

Table 5.2. *The Perfectly Controlled Experimental Design*

	Treatment	Other Factors (K_i)	
Experimental group	X	Y_2	Constant
Control group		Y_4	Constant

class $= x$."[21] In order to construct such a hypothesis (i.e., to specify x), no procedure is more reliable than taking the SAT score of each and every student in my class and then calculating the mean. Of course, the data (the individual SAT scores) would be collected *before* the hypothesis has been formulated, and indeed *used to construct* the hypothesis itself. Yet, this is certainly no good reason to deny the inductive support provided by the evidence in favor of the hypothesis. So, predictive success does not seem to be a necessary criterion for confirmation.

The second reason to question the predictive success criterion has to do more directly with the logic of experimental testing. As we saw in Chapter 4, at the core of the experimental approach lies the model of the perfectly controlled experiment. Now, in the perfectly controlled experiment, there is no mention of prediction, independent evidence, or similar notions. The basic model, to recall, is reproduced in Table 5.2.

Here the observations provide measures of Y at the posttest stage of the experiment only. The hypothesis under test is that factor X (the treatment) is a cause of Y. Of course in most experiments, the circumstances are deliberately set up so as to constitute a good test of the hypothesis, which is therefore formulated *before* the evidence becomes available. That's why most experiments satisfy the predictive success requirement. But this is not essential for inductive support, and in certain situations in fact, the predictive success criterion is violated. Take the (rare) case of natural experiments, in which scientists are lucky enough to witness a phenomenon that takes place in circumstances that approximate the perfectly controlled design, except that nature has set the initial conditions for us. When such cases occur, of course, it would be foolish to deny that the evidence strongly supports the hypothesis that, say, X causes Y, just because the hypothesis has been constructed post hoc, on the basis of X's and Y's covariation in (naturally) controlled circumstances.

Notice that the ideal experimental design accounts nicely for the Duhemian intuition concerning risk: a good experiment makes it unlikely

[21] For non-American readers: the Standard Aptitude Test (SAT) is widely used in U.S. higher education to assess students' capacities.

that a causal hypothesis is false given the evidence, because the K_i have been prepared so as to make the inference from evidence to hypothesis maximally strong. The idea is that it would be really unlikely to observe e (*that* particular correlation between X and Y) if H were wrong (if X were not a cause of Y). But it would be unlikely because the K_i are such that e wouldn't be produced unless H were true, not because some scientist took a risk à la Popper in making the prediction.

Remember that some sort of reasoning about risk is also involved in Popper's arguments about predictive success. But as noticed by Mayo this is not the "right" kind of risk: Popper argues that hypotheses constructed post hoc to fit known evidence have no chance of being refuted by that evidence, whereas what really matters for inductive support – as highlighted by the model of the perfectly controlled experiment – is that *some* hypotheses constructed in such a way would fit the evidence *even if they were false*. In other words, the problem arises only if that kind of evidence is highly likely to be produced regardless of whether H is true or false.

This is exactly the case with the evidence from classic PR experiments. That evidence does not severely test the Karni and Safra hypothesis because the experiment was not designed to minimize the probability of observing that evidence were the intransitivity hypothesis true. Or, equivalently, it is not at all unlikely that the Karni-Safra hypothesis is false, and yet that kind of evidence is observed. In fact, we know perfectly well that the same evidence is also implied by the intransitivity of preferences hypothesis. In contrast, cases like the SAT counterexample show that it is possible to use the existing evidence to support a "tailored" hypothesis independently of the temporal relation between data collection and hypothesis formation. The important point is that alternative interpretations must be controlled for. This is arguably what Rubinstein has in mind when he says that "if some pattern of behavior, *from among an endless number of possibilities*, is discovered in the data ex post, the results are much less informative" (2001, p. 26, emphasis added). Predictive success is just an epiphenomenon of this more fundamental principle of inductive support.

Intuitively (an intuition that I shall not try to make more precise until the next chapter), a key requirement to impose on the relation between H and e is that the probability of observing a certain kind of evidence, assuming some alternative hypothesis is true, is low. Concretely, we need some experiment especially devised to investigate the possibility that violations of independence are to be blamed for the price-choice reversals, something that the original PR experiments were not designed to do. In the next chapter, we shall see how this can be (and has been) done.

Conclusion: The background to the forefront

We can now begin to see what is wrong with predictive success. This criterion focuses on the relation between H and e only – the temporal relation between them.[22] However, it fails to mention a crucial element: the background factors K_i. It is *by virtue of these factors* that the evidence collected in the perfectly controlled experiment strongly supports the hypothesis at stake. This is the third reason to doubt that the predictive success criterion captured the essence of the experimental method – it ignores the auxiliary, or "background," conditions. Some experimental scientists highlight the crucial role played by the background conditions in making reliable scientific inferences. Here's a quote from Davis and Holt's experimental economics textbook, for example:

Tests of market propositions with natural data are joint tests of a rather compli- cated set of primary and auxiliary hypotheses. Unless auxiliary hypotheses are valid, tests of primary hypotheses provide little indisputable information. On the one hand, negative results do not allow rejection of a theory. [. . .] On the other hand, even very supportive results may be misleading because a test may generate the "right" results, but for the wrong reason [. . .]. *Laboratory methods allow a dramatic reduction in the number of auxiliary hypotheses involved in examining a primary hypothesis.* (1993, p. 16, emphasis added)

In the next chapter, I defend a view very similar to the one expressed in Davis and Holt's last sentence: the advantage of controlled experimen- tation is that it allows the elimination and control of the background fac- tors that may confound the inference from evidence to hypothesis. Intu- itively, the advantage of laboratory over nonlaboratory science seems to lie in the fact that the evidence in the former case is collected in "ideal" circumstances, in which we are confident that the background conditions are "right." However, the predictive success requirement does not make any principled distinction between the two cases. We should be suspi- cious of this: if experimental tests are universally considered a privileged source of scientific knowledge, there must be a reason for that, and our criterion of inductive support must be able to capture such a reason.

[22] The same can be said of the construct-independence criterion (see n. 20).

The idea that inductive support is a three-place relation among H, e, and K_i rather than a two-place relation between H and e has some drastic philosophical implications, which partly explains why philosophers of science have been so reluctant to endorse it. The inductivist program, as I mentioned, aimed at doing for inductive inferences what logicians had done for deductive ones. It would be nice if one could provide a set of rules such that by inspecting the purely formal features of H and e, it were possible to establish whether e supports H, and to what extent. Once the K_i enter the picture, however, the issue of inductive support becomes contextualized: one cannot answer it by merely looking at the features of e and H. An empirical investigation is necessary in order to establish whether the context is "right" for e to be truly confirming evidence for H or not. No more can be said on this issue before I illustrate in more detail what role the background conditions play in scientific testing. Here I can just highlight that the role of the philosopher/methodologist and that of the scientist tend to blur under this approach. Philosophy cannot provide a set of rules of inference valid a priori if inductive relations are empirical in character. Scientists' knowledge of the context and circumstances of research is required in order to assess the validity of scientific inferences. Such a conclusion is rather more digestible now than it was half a century ago, because philosophers today generally endorse the view that epistemology should be "naturalized," or partly based on the results of science itself. There are various views about how this can be done, but this is too big an issue to be discussed properly here (cf. Rosenberg 1996 for an overview). My main topic is experimental inference, and the next chapter discusses an approach based on the old-fashioned method of eliminative induction.

SIX

Elimination

The attempt to ground inductive support on the predictive success requirement faces some serious objections. Such a project seems to miss a crucial element in confirmation, namely the background conditions (K_i) that allow us to forge a tight link between the evidence (e) and the hypothesis under test (H). Unless we model the K_i in the relation of inductive support, our theory of induction will be unable to explain the advantages of controlled experimentation over nonexperimental methods of investigation. In experimental science, the K_i can be systematically checked and thus possible sources of error eliminated. In this chapter, I further elaborate this idea and illustrate its virtues using the example of preference reversals.

Subjective Bayesianism

I have said repeatedly that a satisfactory philosophy of experiment should account for the privileged status of controlled experimentation among the various methodologies that scientists use. It might be instructive then to start with an approach that recognizes the importance of background assumptions for induction, but fails to draw a sharp distinction between experimental and nonexperimental evidence. It is the subjective (or "personalist") Bayesian approach, which during the eighties was on the verge of becoming the standard view of inductive inference in philosophy of science.[1] Economists should be familiar with its main characteristics,

---

[1] For a thorough defense of the Bayesian approach to inductive inference, see Howson and Urbach (1989). The original Bayesian solution to Duhem's problem is due to Dorling (1979), whereas Soberg (in press) applies Bayesian analysis to experimental economics.

110

because agents' beliefs and preferences in standard economic models are represented in subjective Bayesian terms. But whatever its virtues for modeling economic agents, this approach is not adequate for a theory of scientific inference.

The key thesis of Bayesianism is that the degree of support provided by a piece of evidence (e) to a given hypothesis (H_1) is measured by a probability value. The posterior probability of a hypothesis given a piece of evidence, then, can be calculated using Bayes's theorem, a theorem of the standard probability calculus:

$$P(H_1 \mid e) = \frac{P(e \mid H_1)P(H_1)}{P(e)}.$$

It is easy to see that when the same evidence can be explained by two competing hypotheses (H_1 and H_2), the posterior probability (and hence the degree of confirmation) of each hypothesis depends entirely on two factors: the likelihoods (the probability of the evidence given a hypothesis) and the prior probability of each hypothesis before the evidence is collected. The posterior probability of H_2, in fact, is given by

$$P(H_2 \mid e) = \frac{P(e \mid H_2)P(H_2)}{P(e)}.$$

Because the two formulas have identical denominators, any difference in the posteriors must be the result of a difference in $P(e \mid H_1)P(H_1)$ versus $P(e \mid H_2)P(H_2)$. Consider a doctor assessing the *positive* outcome of a medical test, e; H_1 is the hypothesis that the patient has a certain condition, H_2 that she hasn't got it. If $P(e \mid H_1)$ is high and $P(e \mid H_2)$ is low – that is, if the test has few false negatives and few false positives – the evidence will have a greater positive impact on the "condition" hypothesis (H_1) than on the "no condition" hypothesis. However, the impact of the likelihoods must be tempered by the prior probabilities $P(H_1)$ and $P(H_2)$: if the condition is very rare in the relevant population, for example, then the posterior probability $P(H_1 \mid e)$ may still turn out to be lower than the alternative $P(H_2 \mid e)$.[2]

When the priors represent *frequencies*, the application of Bayes's theorem may seem to be fairly unproblematic.[3] The problem is that usually

---

[2] For instance: if $P(e \mid H_1) = .8$, $P(e \mid H_2) = .2$, $P(H_1) = .01$, $P(H_2) = .99$, $P(e) = .206$, then $P(H_1 \mid e) = .03$ whereas $P(H_2 \mid e) = .96$.

[3] I say "it may seem" because it is not clear whether the frequency of a condition in a population of different individuals is the right prior to be considered in a case like this. It can be argued that the relevant prior (the probability of *that* particular individual having

in science, our hypotheses do not afford such an interpretation. (What would the frequency of the theory of Quantum Mechanics be? What would it mean to say that such a theory has a frequency in the first place?) Most Bayesians, then, try to save their program by endorsing a *subjective interpretation* of probability claims, as propositions expressing *degrees of beliefs* in a hypothesis. In a way, this is equivalent to recognizing that necessarily we face every decision-situation loaded with prejudice. So-called subjective Bayesians bite the bullet and illustrate how prejudice influences decision making and the updating of our beliefs.

Take a simplified Duhemian case with just two elements, H and K. If we could assign a degree of prior probability (to be interpreted as degree of belief *before* the collection of the evidence) to each element, then by applying the rules of probability theory, the decision whether to endorse (or blame) H or K in the light of e would follow automatically. Suppose for instance that we believe that H is much more likely than K. In such a case, it would be consistent to blame K for the predictive failure and to keep H, at least for the time being. In the case of preference reversals (PR for short – see the previous chapter), we would have the following scenario: H is the principle of transitivity; K is the assumption that the BDM and RLS mechanisms do not lead us into error in observing reversals. The decision whether to take transitivity as falsified would depend on our prior confidence in transitivity as well as on the prior degree of reliability assigned to the elicitation procedures.

It is a consequence of Bayes's theorem that

$$\frac{P(H \& K_1 \mid e)}{P(\sim H \& K_2 \mid e)} = \frac{P(e \mid H \& K_1)}{P(e \mid \sim H \& K_2)} \frac{P(H \& K_1)}{P(\sim H \& K_2)}.$$

Now consider that by using suitable auxiliary and background assumptions, it is always possible to make sure that the likelihoods $P(e \mid H \& K_1)$ and $P(e \mid \sim H \& K_2)$ are both equal or very close to one. If the main hypothesis entails the evidence, if we postulate certain values for the free parameters as Karni and Safra did and we assign an appropriate degree of belief in the working of the apparatus, for instance, then the likelihood of the evidence given $H \& K_1$ may be set so as to be very high (Salmon 1990 calls this

---

that condition) is the propensity associated with the genetic, environmental, etc. factors of *that* particular individual. The relevant chance setup, in other words, is not the one generating the frequency of the condition in the overall population (because there is no such unique setup), but the conditions in which *that* specific individual happens to be. Of course, we very rarely have access to such information, which according to some philosopher is why we need a method of inductive inference that does not rely on priors. See also the discussion between Howson (1997a, 1997b) and Mayo (1997a, 1997b).

procedure "the invention of plausible scenarios"). The same, of course, can be done for $\sim H$ & K_2. However, if $P(e \mid H$ & $K_2) = P(e \mid \sim H$ & $K_1) -$ if the observation of reversals, for example, is entailed by the assumption that transitivity is true and the elicitation procedures false, as well as by its opposite – then the impact of the evidence on each alternative ("H & K_1" vs. "$\sim H$ & K_2") depends entirely on the prior probabilities assigned to the alternatives themselves (i.e., the ratio $P(H$ & $K_1)/P(\sim H$ & $K_2)$). If, for example, $P(H$ & $K_1) > P(\sim H$ & $K_2)$, then $P(H$ & $K_1 \mid e) > P(\sim H$ & $K_2 \mid e)$.

The probability calculus, in other words, can be used to demonstrate that *if* you believe so and so *then* you must believe so and so, or that certain posterior beliefs must follow rationally from certain prior beliefs and the evidence. However, this is slightly disturbing: are we ready to accept *any* degree of prior belief, no matter how crazy or unsupported? As empiricists, we want objective empirical reasons to be embodied in scientific methodology. Bayesians reply that their approach *is* empiricist in character, indeed that it captures all the rationality empiricism can afford. Evidence, they argue, does always have an impact on our beliefs, albeit in an indirect way. Consider the "losing" alternative in the example above, $\sim H$ & K_2. Although the impact of the evidence is greater on its rival (H & K_1), $\sim H$ & K_2 will also be supported to a certain (perhaps minimal) degree. As more evidence of this kind is collected, the probability of $\sim H$ & K_2 will grow, and will eventually become very high.[4] The priors, as Bayesians like to point out, are "washed out" by the evidence and eventually, in the long run, do not count. But in the meantime, they affect the way in which different scientists interpret the same evidence.

Superficially, this seems to be consistent with the history of many scientific debates, including the one on PR: few economists abandoned right away the standard theory of choice in the light of Lichtenstein and Slovic's results, and those who did were probably those who already questioned the theory anyway. Most neoclassical economists, in contrast, were skeptical of the PR result and stuck to their favorite models of rational behavior. According to subjective Bayesianism, this move was (subjectively) rational – and so was the attitude of those, like Lichtenstein and Slovic, who took the PR result as refuting standard choice theory. Eventually, after enough experimentation, both types of prejudice should wash out, but in the meantime, there is no way to tell who's right or wrong.

---

[4] See, e.g., Redhead (1980) for a demonstration of how the probability of a hypothesis can be raised by testing it repeatedly in conjunction with different background assumptions.

However, there is an important sense in which the dogmatic economist who assigns a prior probability of 0.00001 to the hypothesis that the transitivity axiom is false, and dismisses the PR phenomenon *for this reason only*, is irrational, for she has no good empirical reason to do so.[5] We need a methodology that is able to recognize and blame prior beliefs of this sort as irrational prejudices, instead of merely describing them. In fact, if we look more carefully at the PR controversy, we realize that the Bayesian story is not entirely convincing. True, scientists have prior beliefs (although whether they are expressed in probability measures is doubtful); and true, they don't give them up easily: they tend to stick to their guns until more evidence is brought to bear on the issue at stake. We may also grant that their prejudice is a cause of their conservatism. However, it seems bizarre to say that it is *rational* not to believe in the implications of a certain experimental result just because one has a certain prejudice. Whether an individual scientist or group of scientists is strongly prejudiced for or against a given hypothesis is a useful piece of information in order to understand why they try to prove or disprove that hypothesis, but should not determine whether a piece of evidence does or does not disprove it, as a matter of fact. And, in fact, no scientist will ever say: I don't believe in PR because I don't want to give up the principle of transitivity. What they do say is: I don't believe in PR because I have good reasons to suspect that the PR experiment may be flawed. Only in the latter case do they feel rationally entitled to disbelieve an experimental result, because they distinguish correctly between subjective beliefs and objective matters of fact.

Take an example from physics. In 1919, Eddington argued that his measurements of the deflection of light near the sun refuted Newton's theory and confirmed Einstein's predictions. The problem was that not all observations collected by Eddington were consistent with Einstein's predictions, an inconsistency that Eddington resolved by blaming one of his instruments of observation (a mirror used to observe the eclipse). However, many other conservative scientists questioned Eddington's explanation in an attempt to save Newton's theory of gravitation. Eventually Eddington's pro-Einstein party won, after a good deal of argument over the analysis of data.[6] As a prominent critic of Bayesianism puts it,

---

[5] An even more difficult case for the personalist is one in which a scientist assigns a zero prior probability to a hypothesis: in this case, no evidence will ever be able to raise its posterior probability.

[6] For two different accounts of this controversy, see Earman and Glymour (1980) and Collins and Pinch (1993, Ch. 2).

Eddington believed in the correctness of Einstein's account, but nobody cared how strongly Eddington believed in Einstein. Quite the contrary – it only made those who favoured a Newtonian explanation that much more suspicious of Eddington's suggestion that the faulty mirror, not Einstein's account, was to blame. (Mayo 1996, p. 110)

Similarly, in the PR case, we need an objectively good reason to design new experiments to test the alternative explanations of price-choice reversals; it is not enough to say that the refusal to accept this result was due to economists' prejudice.[7] (We should leave such explanations to historians, who are usually much better than philosophers at that.) Finally, notice that subjective Bayesianism fails once again to distinguish appropriately between experimental and nonexperimental evidence. And the reason it fails is strictly related to the issue of prior beliefs. Although Bayesians are able to incorporate background assumptions in their calculus, they do not provide any grounds to treat differently background assumptions in experimental and in nonexperimental science. What matters for them is not scientists' actual degree of control over background factors, but their prior degrees of belief, which may be set in a totally arbitrary fashion.

Objective inductive support

The impact of a given piece of evidence on a scientific hypothesis should not depend merely on what individual scientists think of it. Following Peter Achinstein (2001), it is useful to distinguish between two separate issues here: whether a scientist does (or ought to) believe that *e* supports *H*, given her overall system of beliefs; and whether *e* does support *H* independently of what that scientist thinks, knows, or believes. We may say that a theory aiming at resolving the former issue is a *subjective* theory of induction, whereas a theory focusing on the latter is concerned with the *objective* relation between *H* and *e*.[8]

---

[7] For more detailed critiques of the subjective Bayesian approach to Duhem's problem, see Earman (1992), Worrall (1993), and Mayo (1996, especially Ch. 4). On economists' dogmatic attitude toward preference reversals, see in particular Hausman (1992a, Ch. 13) and Hausman and Mongin (1998). Tammi (1999) reconstructs the debate on PR as a series of "escape moves" aimed at limiting the impact of the anomaly on the received theory.

[8] Achinstein (2001) draws a more complex taxonomy of confirmation theories by distinguishing among four concepts of evidence: "subjective," "epistemic," "potential," and "veridical." Roughly, the first two correspond to my "subjective" approach, the latter two to my "objective" approach. My simpler (and rougher) classification is good enough for the purposes of this book. For a similar distinction ("actual" vs. "assessed" support), see Lipton (1991, p. 151).

Desubjectivizing induction, however, is not easy. In light of what was said in the previous section, it may seem promising to start by just taking away the problematic elements from Bayes's theorem – the priors. What we are left with are the likelihoods, statements that enjoy a slightly more respectable status, from an objectivist point of view. It may seem natural, then, to propose the following criterion: a piece of evidence e supports or confirms a hypothesis H_1 more strongly than a rival H_2 if and only if $P(e \mid H_1) > P(e \mid H_2)$. This "law of likelihood," as it has been called, provides at least a comparative principle of inductive support.[9] Moreover, it is a weak principle that is not inconsistent with the logic of Bayes's theorem: by itself, the law of likelihood is agnostic on what the exact posterior probabilities of the two competing hypotheses will be; the law is merely concerned with the relative *impact* of the evidence on each hypothesis. Bayesians are then free to "temper" the impact of the likelihoods using whatever prior probabilities they think can be legitimately used.

Likelihoodism, however, must face some powerful objections. The first problem comes from "concocted" hypotheses, that is, hypotheses especially created so as to make likelihoods maximally high. We already have seen an example in the previous section, and again the PR debate provides a concrete instantiation of this problem: by adding specially crafted background and auxiliary assumptions K_i, it is always possible to make sure that the evidence is logically *implied* by the "cluster" of hypotheses and auxiliaries, as first highlighted by Duhem and Quine. However, these concocted hypotheses as a consequence will always receive maximum support from the evidence, contrary to our intuition.

There are, to be sure, some technical problems in defining the likelihood values of composite hypotheses like the ones just mentioned.[10] To avoid technicalities, however, consider the simple hypothesis $H_3 = $ "an evil demon made the patient test positive for the condition." No matter what the alternative is, this hypothesis will receive at least as much support from the evidence as its rivals – which is, of course, rather disturbing. One possible strategy is again an appeal to modesty; Elliot Sober, a prominent likelihoodist, puts it as follows:

Likelihoodists can and should admit that the demon hypothesis is implausible or absurd, notwithstanding the fact that it has a likelihood of unity [. . .]. It's just that likelihoodists decline to represent this epistemic judgment by assigning the hypothesis a probability. Likelihoodist epistemology is modest in its ambitions; support gets represented formally, but plausibility does not. (Sober 2002, p. 24)

[9] Hacking (1965) provides a classic statement of the law of likelihood. For a more recent discussion of "likelihoodism," see Forster and Sober (in press).
[10] Cf. e.g., Sober (2002) for a brief illustration.

Reasonable as this suggestion might sound, it leaves open the question of how to justify in a nonsubjective fashion the implausibility of the demon hypothesis. And even if we could do that, this would leave us with the rather unpleasant fact that the evidence *does* support H_3 better than the alternatives, although H_3 is ruled out on independent (albeit unspecified) grounds of plausibility. However, this is not very sensible: surely what we want to say is that H_3 does not receive any support from the evidence, because it is not the kind of hypothesis that can be checked by means of a medical test.

To articulate this intuition, a little "gestalt shift" is required. Remember that according to Bayesianism confirmation is to be represented by means of a probability measure. Ronald Giere (1977) has proposed to label the theories that assign such a role to probability as *information models* of inductive inference.[11] The contrast class is provided by so-called *testing models* of inductive inference – theories that use probabilities to measure the properties of testing procedures (e.g., the power or the severity of a test) instead of the degree of confirmation of scientific hypotheses. Well-known examples include orthodox (Neyman-Pearson) statistics, Popper's falsificationism, and, more recently, Deborah Mayo's error-probabilistic account.

We shall see later how to apply a very general version of the testing approach to experimental reasoning, but for the time being, let me highlight a couple of key differences with respect to the likelihoodist and Bayesian approaches. What we want is to impose a condition that links support or confirmation not only to the high probability of observing the evidence if the hypothesis is true (as in likelihoodism), but also to the absence or the low probability of alternative explanations of the evidence. It is precisely because this latter condition is satisfied that the patient should be worried by the positive result of a test with low false positives – because the probability of observing that result, if the condition is absent, is very low. (Readers familiar with orthodox statistical methods should find these notions familiar: standard Neyman-Pearson statistical tests can, in fact, be seen as an attempt to implement them in a more precise and rigorous way.)

Giere (1983) and Mayo (1996) propose that a hypothesis H should be considered indicated (qualitatively) by the evidence e only if the test that produced e is such that there is a high probability of observing e if the hypothesis is true, and a low probability of observing e if it is false.

---

[11] Deborah Mayo's (1996) *evidential relationship* category tries to capture essentially the same class of theories of inductive inference, albeit by means of a different terminology.

This is generally the line that I shall follow in the rest of the chapter, although – this not being a book on induction – I shall try to keep the discussion at a rather general level and will not make any commitment on a number of specific issues of inductive inference. My main effort will be devoted to illustrate how such a requirement is translated at the level of experimental design, that is, how experimental scientists create their experiments exactly with this requirement in mind: they aim, in other words, at designing test situations such that the evidence points unequivocally in the direction of the hypothesis under test.

By introducing the model of the ideal controlled experiment, in a way I have already identified a situation in which the requirement is *objectively* and *perfectly* satisfied. The perfectly controlled experiment, in fact, describes a situation in which a given piece of evidence can be used to make the strongest possible inference to a causal hypothesis. If the evidence is collected in a controlled experiment, the relation between e and H will be strong – in the objective sense – regardless of what scientists may think about it. It will be objectively strong even if no one knows that the evidence has been collected under controlled experimental circumstances. That's why a given piece of evidence may confirm a hypothesis in the objective sense but, at the same time, be seen as not confirming that hypothesis (perhaps even as falsifying it) in the subjective sense.

Another way to put it is this: whether e supports H objectively speaking depends on the K_i, that is, on whether the background circumstances are "right" for that inference to be made. In contrast, whether e supports H in the subjective sense depends on what people *think* of the K_i. A flawed (badly designed) experiment cannot provide evidence for or against a given hypothesis in the objective sense, but may do so in the subjective sense. Now, one may argue that an objective theory of inductive support is of little practical use: after all, in real life, we are never sure that we have got adequate knowledge of the circumstances under which we act. Duhemian problems in real life have a subjective component in them. This is certainly true, and in fact, later in this chapter, I relax the objectivist approach by injecting a social component in the notion of inductive support. However, we should not dismiss the objectivist approach too quickly. To specify a set of circumstances in which the objective support is maximally strong is useful in order to identify what sort of situation must occur for us to have the "right" sort of beliefs. It tells us what we should look for to correctly infer from e to H.

The model of the perfectly controlled experiment also indicates what constitutes a legitimate challenge against an experimental result, and what scientists should do to rebut such challenge, or to make sure that a given experiment has been performed correctly. As just pointed out, a flawed experiment is one in which some background factor "interferes" to create the illusion of a phenomenon that is not really there (an *artifact*); such an experiment clearly does not provide good evidence in the objective sense. Therefore, experimenters should make sure that no mistakes have been made in designing or performing the experimental test, or that all the K_i have been controlled adequately. And in fact, most experimental science is simply devoted to checking that no errors have been made in other, previous experiments – they are aimed at eliminating artifacts by checking the robustness of phenomena and the correctness of their explanations.

Experimental error and checking

A key advantage of controlled experimentation is that it allows the systematic search and elimination of errors. Experimental induction, in this sense, is eliminative induction. This is not a particularly illuminating claim in itself, for as John Earman (1992) rightly points out, all inductive strategies must in one way or another be eliminative in character.[12] Remember the underdetermination problem: logically speaking, a given piece of evidence can be entailed by an infinite number of alternative hypotheses. Hence, inductive inferences must somehow select from the infinite number of possibilities those that are truly supported by the evidence from those that are not. The interesting issue is how this selection takes place. *Pace* Earman, there are reasons to believe that the elimination does not always or typically take place at the level of "grand" theories. Such reasons come, on the one hand, from the observation of actual experimental practice and, on the other, from the analysis of the perfectly controlled experimental design.

The hypotheses at stake in many experiments are low-level claims about the occurrence of certain phenomena, the functioning of a measurement instrument, or the influence of some background factor. Another way to describe such activities is to say that these experiments are aimed at error detection and elimination. The "experimental artifacts" discussed

---

[12] Eliminative induction, traditionally associated with Francis Bacon's *Novum Organum* (1620), has suffered from a lot of bad press in contemporary philosophy of science, but has recently begun to be rehabilitated. Influential defenses of eliminative induction include Mackie (1974, especially the Appendix), Earman (1992, Ch. 7), and Kitcher (1993, Ch. 7).

in previous chapters are exactly this: errors in drawing inferences from data to phenomena, from phenomena to theories, or from what happens in an experimental setting to what is going on in nonlaboratory conditions. Scientists sometimes test the claim that a phenomenon is real, sometimes check the accuracy and reliability of instruments, and anyway try to make sure that the data they collect constitute good evidence for whatever purpose they are supposed to be used. All this may or may not be related to the testing of some high-level explanatory theory, eventually. However, what is eliminated in each instance usually is not a high-level alternative explanation of the evidence, but rather a possible source of experimental error that may confound the inference from the evidence to the hypothesis of interest.

Thus, for example, most experiments on PRs tested the hypothesis that PRs are a real phenomenon rather than an artifact of the instruments of preference elicitation. The elimination took place at the level of auxiliary assumptions such as "the BDM procedure is flawed," or "the incentives are not adequate." Theory testing on a grand scale is an activity that can at best be carried out within a whole *research program*, that is, in a long series of related experiments. Within a research program, each single experiment is typically devoted to low-level hypothesis testing. This is not a new thesis, for several historians and philosophers of science who have carefully studied experimental practice have made a similar point before. The physicist Allan Franklin (1986, 1990, 1998), the historian Peter Galison (1987), and the philosophers Giora Hon (1989) and Deborah Mayo (1996) are among those who have defended this view with most vigor recently.

The second argument in favor of eliminative induction "in the background" comes from the perfectly controlled experimental design. This model experimental design highlights that, contrary to what some overenthusiasts of underdetermination suggest, (1) inferences from the evidence can in principle be local, that is, can be directed toward a well-specified hypothesis in the "cluster" of all explanations that are logically compatible with the evidence (typically, that a certain factor X does or does not causally influence another factor Y). And furthermore, that (2) the reliability of such local inferences depends largely on the control of background factors. Such background factors are the aspects of the design that can mess up the experimental inference or create artifacts. An objectively good experiment, then, is such that the known possible sources of error have been eliminated *ex ante* by means of accurate experimental design.

Notice that the Duhem-Quine problem, as seen from this perspective, takes a distinctively constructive flavor. Many philosophers tend to see

Duhem's problem as a powerful weapon in the hands of the skeptics or the conservatives: it is always possible, according to this interpretation, to question the validity of a result (or to resist an apparent refutation) by blaming some background assumption. For example, here's a typical quote from Imre Lakatos:

No experimental result can ever kill a theory: any theory can be saved from counterinstances either by some auxiliary hypotheses or by a suitable reinterpretation of its terms. (1970, p. 32)

Quine suggests as much in his famous (1953) essay. And Vernon Smith also follows Lakatos and Quine in a recent paper:

The interpretation of observations in relation to a theoretical hypothesis is inherently and inescapably ambiguous, contrary to our [i.e., economists'] accustomed thinking and rhetoric. (2002, p. 98)

The truth, of course, is that the interpretation of data appears "ambiguous" only if we take purely deductive logic as our standard of clarity. However, that is a dubious assumption: we can surely imagine situations (such as the perfectly controlled experiment) in which the evidence makes one hypothesis unambiguously likely, given the background assumptions. Moreover, in many cases, Duhemian arguments are more accurately characterized as instances of constructive criticism than as skeptical challenges. To indicate a potential flaw in an experiment is functional to devising a new control on that particular source of error. Of course, this is not to deny that scientists are, in many cases, motivated by a priori skepticism or conservatism. (As we have seen, the PR case may be exactly an instance of this sort.) Rather, it means that we should distinguish between scientists' psychological motivations and the objective process of scientific research. Whatever the motivations behind economists' critique of the PR experiments, the repeated and ingenious checks prompted by their challenging arguments have strengthened the PR phenomenon, instead of weakening it. Once all the main critiques to the PR design have been met, experimenters simply "do not see how to make the phenomenon go away," to use a physicist's eloquent expression (Galison 1987, p. 235), and accept it as real.

Checking the preference reversals experiment

We left the PR debate right after the formulation of Holt's, Karni and Safra's, and Segal's critiques. These authors, to recap briefly, suggested in a series of independent papers that the PR "phenomenon" may be an

artifact of the instruments of observation (preference elicitation) used
by Lichtenstein and Slovic, and Grether and Plott. This case provides a
severe test for the view that inferences within the experiment are eliminativist in character, because of the highly theoretical character of the
arguments proposed to discredit the PR experiments. It is tempting to
interpret the controversy as a contest among alternative theories, which
is eventually settled by drawing predictions from each theory and testing
them in the laboratory. However, as I said, the PR debate is best seen as a
series of attempts to discover and check potential artifacts in the original
experiment devised by Lichtenstein and Slovic. This appears obvious if
one looks at experiments such as Grether and Plott's (1979), but the same
can be said of later empirical work in this area.

I have argued in Chapter 5 that the alternative explanation of the PR
data (the "price-choice reversals") proposed by Karni and Safra could not
receive inductive support from the original PR experiments because the
latter had not been conceived with the aim of controlling for violations of
independence. Still, the Karni-Safra critique indicated a potential flaw in
the PR experiment, an error that could affect the inference from the data
to the PR phenomenon (to the intransitivity of the preference relation, in
particular). Some new experiment had to be specially designed to check
this potential source of error. In fact, we find in the experimental literature
of the late eighties and mid-nineties a number of tests devised for this
purpose. Here I briefly illustrate just a few representative examples.

The first test I want to examine looks for signals of the experimental
flaw. "Disturbing" factors and other sources of error (e.g., malfunctioning
apparatus) typically leave traces that can be detected under the appropriate conditions. A checking experiment simply creates such conditions.
In our case, the experiment tests the joint effect of reduction and violations of independence. Starmer and Sugden (1991) have tried to create
experimental circumstances under which the reduction hypothesis, upon
which critiques to the BDM procedures are built, is incompatible with a
very frequent effect, a violation of independence first discovered by Maurice Allais (1953) and known as "common consequence." The experiment
involves a double choice, first between a lottery $R' = (£10, 0.2; £7, 0.75; 0, 0.5)$ and a lottery $S' = (£7, 1)$; then between $R'' = (0, 0.8; £10, 0.2)$ and
$S'' = (£7, 0.25; 0, 0.75)$. The common consequence effect is a tendency
to choose $S' \succ R'$ and $R'' \succ S''$: subjects seem to weigh the probabilities
differently depending on whether they provide certainty (as in S') or not
(as in S''). By reduction, it is easy to show that the following equivalence
holds between compound lotteries: $(R', 0.5; S'', 0.5) = (S', 0.5; R'', 0.5) =$

(£10, 0.1; £7, 0.5; 0, 0.4). If there is reduction, then, one should expect a random pattern of choice between (R', 0.5; S'', 0.5) and (S', 0.5; R'', 0.5), whereas common consequence implies (S', 0.5; R'', 0.5) $\succ$ (R', 0.5; S'', 0.5). If there is reduction, in other words, there cannot be common consequence effects, and vice versa. Starmer and Sugden performed the above Allais-type experiment with and without the RLS mechanism, and observed the same ratio of common consequence violations in all cases. This provided evidence that the reduction hypothesis cannot be right, contrary to what some critics of PR experiments had suggested.

Similarly, Safra, Segal, and Spivak (1990a, 1990b) derived from Karni and Safra's model an implication that could be tested in the laboratory. Their "Proposition 2" states that although the optimal selling price (π) of a lottery and its certainty equivalent (CE) may end up being nonidentical in a BDM elicitation (hence the "error" in observing preferences), they should nevertheless lie on the same side of the lottery's expected value (EV). More precisely, the two following testable predictions (for risk-loving and risk-averse subjects, respectively) can be derived from Karni and Safra's interpretation:[13]

   (i) $CE(X) > EV(X) \rightarrow \pi(X) \geq EV(X)$
   (ii) $CE(X) < EV(X) \rightarrow \pi(X) \leq EV(X)$[14]

An experiment testing such predictions would act as a sort of "detector" for the error indicated by Karni and Safra.[15] Keller, Segal, and Wang (1993) ran such an experiment, and found Proposition 2 to be inconsistent with about thirty percent of the data.

Consider that some experimental checks can be conducted *after* the collection of the evidence, in apparent violation of the predictive success criterion. Violations of Proposition 2, for instance, could be interpreted as evidence of occasional random mistakes. Subjects, according to such an interpretation, would have a tendency to act in accordance with the Karni-Safra model but, at the same time, deviate stochastically from the

[13] For the technical details of such a derivation, cf. Safra, Segal, and Spivak (1990b, pp. 187–8).
[14] According to Segal's (1988) interpretation of the BDM device, on the other hand, (i) and (ii) do not necessarily hold. A test of such hypotheses may therefore be seen as a sort of "crucial experiment" discriminating between the Karni-Safra and Segal arguments.
[15] To be more precise: the absence of the two effects above would be a sign of the incorrectness of the Karni-Safra account, whereas the observation of (i) and (ii) would not necessarily count in its favor. The two effects (i) and (ii) can, in fact, be derived also from EUT, because according to the latter, $CE(X) = \pi(X)$.

central tendency. However, a quick look at the evidence will reveal that this cannot be the case. The data display a definite asymmetry: the $\pi(X) >$ $EV(X) > CE(X)$ pattern is displayed for 22 percent of the subjects, whereas the $CE(X) > EV(X) > \pi(X)$ pattern is shown for only 9 percent (Camerer 1995, p. 659). The "random mistake" interpretation can therefore be dismissed with a high degree of confidence.

At this stage, one may be tempted to conclude that these experiments fit perfectly well a theory-testing account: a theory (EUT) is proposed, predictions are derived from it, the evidence (PR) refutes it, another theory is proposed (Karni and Safra's), a prediction is derived from it, evidence refutes it, and so on and on. However, this would be too quick a conclusion. Recall that the sticky issues in the experiments on PR did not primarily have to do with any highly explanatory theory. They can rather be summarized in the question, How can we know that the observations of preference relations made via elicitation mechanisms were reliable? Unfortunately, many philosophers and scientists tend to conflate these two issues: (1) Have we got a correct theory of the instruments of observation? and (2) Do the instruments observe accurately? This confusion tends to generate unnecessary problems and exaggerated worries. In fact, it is not the case that if one does not have a theory of the instruments, one is not entitled to believe in what the instruments show.

Karni and Safra, in their paper devoted to challenging the BDM elicitation mechanism, pose the following two questions: (a) "How rich is the class of preferences that permits the elicitation of certainty equivalents of given lotteries using [the BDM] method?" and (b) "Are there experiments that enable the elicitation of the certainty equivalents of every lottery for every reasonable preference relation?" The first question is the one that motivates their enterprise; the second one is obviously more ambitious, asking as it does for an elicitation procedure that is absolutely general in its scope of application. Of course, they are both legitimate questions, and indeed interesting ones from a scientific (and particularly *theoretical*) point of view, but their relevance to the issue at stake (i.e., the artifactual nature of PR) is far from clear. Karni and Safra answer, respectively, that

(a) the elicitation of certainty equivalents of all lotteries, using the experimental methods of Becker, DeGroot, and Marschak, is possible if and only if the preference relation is representable by an expected utility functional; (b) every experiment in a larger class of experiments [which Karni and Safra call "Q-experiments"] would fail to elicit the certainty equivalent of some lotteries for some reasonable preference relations. (Karni and Safra 1987, p. 676).

In other words: if subjects' choices violate independence, then the BDM procedure and similar mechanisms are not adequate instruments to determine certainty equivalents *in a precise way and in all cases*. From this, Karni and Safra conclude that "Grether and Plott and others [...], as our discussion indicates failed to [...] observe by means of an experimental method developed by Becker, DeGroot, and Marschak (1964), the certainty equivalents of given lotteries" (ibid.). Apart from the fact that – given what we have seen so far, and as later research on PRs has shown – this is quite clearly an overstatement, the point is that such an argument is not in itself sufficient to challenge the existence of PRs. To begin with, the first premise (that agents do actually violate independence in these particular cases) had not been established. Secondly, even if it had been established, it does not follow logically that PRs cannot be observed by means of "Q-experiments." The BDM mechanism and similar methods, in fact, may not be *absolutely* or *generally* precise, but still *precise enough* to observe PRs. One natural way to see whether this is the case or not is to try to observe reversals *with and without* the BDM procedure, and check whether it makes any difference.

Remember the target of Holt's, Karni and Safra's, and Segal's arguments: originally, they intended to show that it was not intransitive preferences that experimentalists had observed in their experiments with the BDM procedure. In order to reject their interpretation, therefore, one need not necessarily show that their alternative accounts are mistaken. It should be sufficient to show that it was a genuine feature of preferences that was observed in the experiments in question. Ian Hacking (1983, Ch. 11) argues that our confidence in what we see through *electron* microscopes is enhanced by the fact that the same entities or phenomena are observed through light microscopes. This procedure is analogous to what social scientists call *triangulation*: if you are not sure about any measurement technique, use several different procedures. The intuition behind this inference is captured by a so-called no-miracles argument:

Two physical processes – electron transmission and fluorescent re-emission – are used to detect the bodies. These processes have virtually nothing in common between them. They are essentially unrelated chunks of physics. It would be a preposterous coincidence if, time and again, two completely different physical processes produced identical visual configurations which were, however, artifacts of the physical processes rather than real structures in the cell. (Hacking 1983, p. 201)

According to such an argument, evidence obtained via independently working instruments provides strong support for the existence of a

phenomenon. Our belief in the reality of a phenomenon can be independent of the explanations of how we were able to observe it. We may
not know the causes of the phenomenon or have an established theory of
the instrument, and yet we may believe in the phenomenon and in what
we detect using that instrument.

Consider a famous example from physics, Jean Perrin's determination
of Avogadro's number. Avogadro's number is the number of molecules in
a mole of any substance. Perrin was able to measure it observing the random movement of tiny particles suspended in a fluid (a phenomenon
known as Brownian motion). In order to be sure that he had found
the true quantity he was after, however, Perrin checked his result by
comparing it with alternative measures. In his 1913 book, *Les atomes*,
Perrin reports thirteen different independent methods to ascertain
Avogadro's number. The "miraculous" convergence of all measures
is taken to be extremely strong proof that the result obtained was
not an artifact of the procedures he had used.[16] As Franklin (1986,
pp. 131–5) points out, exactly for the same reason, the *Review of Particle Properties* provides detailed information about the different devices
used for measurement purposes (automatic spark chambers, counters,
electronic combinations, emulsions, hydrogen bubble chambers, missing-
mass spectrometer, xenon bubble chambers, cloud chambers, propane
bubble chambers, spark chambers, wire chambers, bubble chamber plus
electronics, freon bubble chambers, etc.). The reliability of a measurement
is a function of the number of different techniques delivering consistent
results.

In the case of PRs, we find several attempts to apply the logic of triangulation. Interestingly, the PR phenomenon had been observed right
from the beginning with and without elicitation mechanisms. Among
Lichtenstein and Slovic's early tests (1971), only two involved the BDM
procedure, but reversals were produced in all of them. This fact should
have already been a puzzle to the Holt, Segal, Karni and Safra explanations but for some reason, was overlooked by the economists investigating
the BDM and RLS mechanisms.[17] Years later, Jim Cox and Seth Epstein,
two economists at the Arizona Experimental Lab and De Paul University,

---

[16] Even a conventionalist like Poincaré was struck by such a result; see Nye (1972) for the
full story. Cartwright (1983), Salmon (1984), Mayo (1996), and Achinstein (2001) provide
slightly different philosophical reconstructions of Perrin's experiments.
[17] A possible explanation is that economists generally dismiss experiments without monetary incentives as unreliable and therefore ignored part of the original data presented by
Lichtenstein and Slovic. In Chapter 11, I discuss the issue of incentives in more depth.

looked for a way to reproduce preference reversals using incentive mechanisms but avoiding possible problems with the BDM procedure:

[...] it was necessary that we not use the BDM price elicitation procedure. Furthermore, we concluded that Karni and Safra's Theorem 2 makes it highly unlikely that anyone will be able to design a price elicitation mechanism for choices in a lottery space that does not require the independence axiom. Therefore, we concluded that it would be impossible for us to elicit true selling prices in an experiment that is designed in such a way that behavioral inconsistencies with the independence axiom are not confounded with more fundamental inconsistencies with decision theory. But preference reversals are inherently properties of inconsistent orderings. The absolute magnitude of prices is basically irrelevant; it is the fact that the less preferred lottery is given a higher price that represents an inconsistency with decision theory. (Cox and Epstein 1989, p. 412)

Cox and Epstein tried to design an incentive procedure able to elicit orderings without creating compound lotteries. First, they asked subjects to state their lowest selling price for both lotteries in each pair; the lottery with the lower price was then paid a fixed sum, whereas the other was played out for money. The prices were then compared with subjects' pairwise choices on lotteries obtained by reducing the payoffs of the original lotteries by the announced selling price (so that the probability distribution of returns was kept constant). The procedure is problematic because – as Cox and Epstein (1989, p. 422) admit – the subjects might interpret the pricing task as a choice task. Some critics (e.g., Hausman 1992a, p. 139), indeed, suggest that Cox and Epstein's might not even be classifiable as a genuine PR experiment. For our purposes, however, the general strategy is what matters: Cox and Epstein's procedure does not prevent the subjects from stating a higher selling price than their true reservation for the preferred lottery, but is supposed to ensure that the latter is assigned the highest price anyway. Because of the structure of the experiment, it was not possible to control for wealth effects and for portfolio effects at the same time; Cox and Epstein decided to control for the latter and then cope with wealth effects by means of data analysis. PRs were observed; the pattern of reversals, however, was quite different from that of classic PR experiments. Many unpredicted reversals and fewer predicted asymmetries occurred, thus warranting the suspicion of (at least some element of) random choice behavior.

Cox and Epstein's approach, however, seemed promising. Tversky, Slovic, and Kahneman (1990) later devised an incentive mechanism with the aim of improving on Cox and Epstein's procedure. The basic idea, again, was that ordering rather than the elicitation of true selling prices is what matters for the observation of the PR phenomenon. Rather than doing things concurrently, subjects were *first* asked to price the lotteries in each pair separately, and *then* were faced with the choice task. Subjects were told that only one lottery among the highest priced and the chosen one would have been randomly selected and played. One can attempt to explain away the observed reversals by means of a generalized expected utility model assuming that subjects implement a mixed strategy, that is, by supposing that agents prefer a 50 percent chance to play either the highest or the lowest valued lottery to the option of playing one of them for sure. Such an explanation, however, cannot account for systematic patterns such as those observed in classic PR experiments and replicated by Tversky, Slovic, and Kahneman.

The importance of the background

These experiments constitute just a fraction of the work done on the PR phenomenon during the last two decades. For reasons of space, I cannot discuss others that are equally interesting and important. Collectively, at any rate, these experiments have convinced experimental economists that PRs are likely to be a genuine experimental phenomenon. In his survey in the *Handbook of Experimental Economics*, Colin Camerer (1995, p. 659) concludes that in the light of the evidence accumulated thus far, the PR phenomenon can hardly be considered an artifact of the instruments of observation. He claims that the doubts concerning the BDM and RLS mechanisms that had been raised in the eighties are not vindicated by the experimental evidence, and that anyway the PR phenomenon is observable with and without these elicitation procedures. Experimental economists may not know exactly how the elicitation mechanisms work (although they surely have a better understanding now than twenty years ago), but they are confident that they may be used to observe PRs.

The increased confidence of experimental economists derives from the numerous, repeated checking procedures implemented to control for possible mistakes in the original PR experiment. An experimental inference, in other words, is just as strong as our capacity to control for the factors that may disrupt the inference *itself*. Identifying such potential mistakes – as Holt, Segal, and Karni and Safra did – is functional to the progress of experimental knowledge; it is indeed the first step toward a new experiment aimed at checking for these errors. That's why experimental knowledge grows progressively and slowly, via a series of laboratory tests carried out in the course of many years or even decades. Testing for a particular kind of artifact will typically require a special experimental design, different from the ones implemented until then – hence the systematic variation that is characteristic of experimental science. Notice that each experiment does not necessarily control for just *one* potential artifact at a time. Several potential sources of error can be identified in advance and taken care of in a single experiment. Indeed, a good experiment is one that has been designed so as to eliminate the possibility of several mistakes at once. In contrast, "bad" experiments that do not control even for the most obvious potential sources of error usually do not make it to the stage of publication. A bad paper is usually rejected by referees if the experiment it describes (with *that* particular design, with *those* background conditions) is such that it does not allow a strong inference from the collected evidence to the hypothesis at stake.

This point can be generalized as follows: the significance of a given piece of evidence, *e*, for a given hypothesis, *H*, depends not just on *H* and *e*, but also on the background circumstances under which *e* has been collected. As anticipated at the end of the last chapter, this marks a crucial difference between inductive and deductive inferences. Whether a certain proposition can be *deductively* inferred from another proposition or set of propositions depends on these propositions only. It does not matter whether certain other conditions (or propositions) hold "in the background." To check that a deductive inference is valid, therefore, we just have to look at the truth-conditions of the premises and of the conclusion. (We do not even have to look at the actual truth of these propositions, for the validity of an inference is a matter that can be settled a priori.) In contrast, the strength of an inductive inference cannot be established independently of the relevant background circumstances: in order to know whether *e* supports *H* or not, scientists have to collect a lot of empirical evidence about the context in which *e* has been generated or observed. As many philosophers have pointed out, the evidential relation is a posteriori (cf. e.g., Sober 1988, Ch. 2; Achinstein 2001; Norton 2003). That's why, incidentally, a lot of experience and detailed knowledge of the subject matter is required to construct a good experiment. And this fact also explains why textbooks of experimental science usually include not only general methodological principles, but also a good deal of concrete information about the specific entities one intends to experiment upon (cf. for instance, Davis and Holt 1993 or Friedman and Sunder 1994). You cannot learn how to become a good experimental economist from a philosophy of science book.

Notice that if inductive relations are empirical in character, they may well be mistaken. Unknown to us, the K_i may not be such as to support a strong objective inference from e to H. This is, of course, just another way of saying that the experimental method is fallible – it is the strongest method we have, but it does not provide absolute certainty. The method is only as strong as our capacity to control the background factors and to design conditions under which we can answer the specific questions we are interested in. Critics of empiricism like to point out at this stage that our knowledge "lacks foundations," or that any scientific claim sits on a swamp of assumptions rather than on a firm bedrock of established facts. In some sense, this is obviously true, but it is important to appreciate exactly what the implications are for empiricism.

The "other factors" problem

The most challenging objection to the project of grounding scientific knowledge on eliminative induction has to do with the open-endedness of eliminative procedures. One of the things that you learn when you are trained as an experimenter is the list of likely artifacts or flaws that are worth checking in a given experimental context. This list is part of what Thomas Kuhn (1962) called the "paradigm": a set of rules to do good normal science. The list is partly determined by purely social factors: experimental economists, for instance, are taught to worry about slightly different problems than those of economic psychologists (incentives and deception, for instance, are a well-known point of contention). Likely sources of error include the instruments of observation, the initial and boundary conditions, but also the presence or absence of many interfering or disturbing factors. Some of these will be known, but others may not be. Scientists (and economists in particular) usually say that a certain conclusion follows from a given set of assumptions or a model only if a ceteris paribus condition holds, that is, if the "other factors" in the background are "equal" or (more appropriately) "absent." There is, then, a sort of indeterminacy in experimental science. At a given moment in time, experimenters may not be able to control for *all possible* flaws, because some of them may not even be known or conceivable given the present state of scientific knowledge.

At this stage, it is worth recalling the distinction between objective and subjective inductive support. Scientists' knowledge is obviously limited, and one can never know for sure whether a result is objectively warranted or not. However, the decision to take a result as confirmed (or disconfirmed) by the available evidence is not up to an *individual* scientist's subjective judgement. Subjective Bayesianism makes the mistake of saying just that, whereas a realistic eliminative approach should recognize that the list of errors and possible flaws is *socially* determined. What scientist *x* thinks about the flaws of a certain design matters relatively little – it is the scientific knowledge of her time (or the community of scientists) that provides a preliminary list of factors that are likely to matter in a given experimental context. An individual scientist, therefore, is to a great extent constrained by the particular historical conditions she happens to live in. She cannot, for example, be blamed for not checking potential mistakes that are not recognized by the science of her time. Whether *e* is interpreted as supporting *H* at time *t* depends on the accepted scientific knowledge at time *t* – and the identification of such a body of knowledge is largely a social rather than an individual matter.[18]

Another Bayesian mistake lies in the requirement that quantitative values be assigned to the priors, which in turn decisively affect the posterior probabilities obtained by means of Bayes's theorem. The eliminativist account does not require quantitative assignments of this kind, not even socially determined ones. All that is needed is the identification of alternative hypotheses and possible sources of error, and the creation of experimental conditions in which they can be appropriately controlled for. To impose on the scientific community the burden of performing Bayesian calculations of posterior probabilities seems as unrealistic as to impose them on individual scientists.[19]

However, despite the reliance of the eliminativist approach on social knowledge, one should not overestimate the importance of paradigms. Strictly speaking, you do not need the full list of possible errors before you start experimenting, because you typically discover many of them as you proceed. (As J. L. Mackie 1974, p. 320, reminds us, part of the purpose of causal analysis is to *find out* about causes.) Experimental scientists are well aware of the fact that some experimental flaws and disturbances may escape their imagination. Thus, a large part of the preparation of an experiment is devoted to meticulous and systematic scrutiny of the design. I describe this activity in Chapter 2, in the sections devoted to the preparation of pilot experiments. Experimenters know that a number of flaws

---

[18] Peter Galison's dictum – "experimental results are accepted when *scientists* don't know how to make them go away" – should always be formulated in the plural, to stress that the relevant unit of analysis is the group or community of scientists, not the individual.

[19] From this respect, my account differs from the "socially oriented" Bayesian approaches of Gillies (1991) and Earman (1992).

are usually discovered during this phase, and that there is no valid substitute to a careful, systematic search. Although we cannot fully imagine the sort of errors we have made in designing the experiment, we can often spot them when we come across them. This activity of careful checking is an invaluable source of confidence that no major flaw is involved in the design and performance of the experiment.[20] Far from being "trapped" in their own presuppositions, scientists are often able to transcend the limits of their own paradigm and thus engage in an effective hunt and elimination of error.

Secondly, scientists working in different disciplines do not live in separate worlds. Many controversies take place across different disciplines, and profitably so. (The PR controversy, of course, is an exemplary case.) The term *scientific community* must be interpreted broadly, to include neighbor disciplines like experimental, social, and cognitive psychology in the case of experimental economics. Interdisciplinarity and the proliferation of alternative paradigms are important because they provide extra challenges to the scientist working in a given tradition. Often scientists in other disciplines have been trained to care for aspects of the experimental design that you are unable to notice or appreciate.[21]

'Other factors' and randomization

However, even interdisciplinary research cannot exhaust the list of possible flaws in an experiment, for the list is potentially infinite. For all we know, it is logically possible that there exists an as yet unobserved factor interfering with some "established" experimental phenomena. If that will turn out to be the case, many experimental results will have to be revised in the light of such a discovery. But how can we know whether this is really the case? One tempting, quick answer is that we do not have to worry about the "other factors" problem because we have experimental techniques to take care of it. Recall that perfect control is not always possible and that in real science, we normally operate in less than ideal circumstances. One technique that is customarily used to cope with imperfect control is *randomization*. Randomization spreads unknown disturbing factors evenly across the treatment and control groups. Thus, for instance, if gender has

[20] Another, similar strategy to deal with the "other factors" problem is to increase the size of the sample of observations, hoping that an unknown disturbing factor will become manifest spontaneously. See Kitcher (1993, pp. 242–7) for examples and discussion.
[21] The point about the importance of interdisciplinarity has been made a long time ago (albeit in slightly different terms) by Paul K. Feyerabend (1975).

Table 6.1. *Additivity*

	No	Low	High
Student	8.0	7.0	4.0
Housewife	6.0	5.0	2.0
Total average	7.0	6.0	3.0

an impact on the phenomenon we are investigating (for reasons that we cannot foresee at this stage of research), we can make sure by randomizing that an approximately equal proportion of male and female subjects are present in each group.

However, this strategy, unfortunately, does not work in all instances. First of all, consider that randomization is usually applied only to certain types of background conditions. Typically, experimenters randomize across subjects and (sometimes) across certain aspects of the design, but not others. They do not, for instance, randomize across the computer terminals' make or the room's temperature. The presumption, of course, is that these factors are not relevant to the result of a typical economic experiment, but clearly this answer just begs the "other factors" question. There surely are many factors that may *in principle* be relevant but that we do not randomize upon.

Moreover, randomization can lead to error in some circumstances. Imagine we are trying to measure the level of contribution in a public goods game in three different conditions (say, high incentives, low incentives, no incentives) in a population, which for simplicity we take to be composed only of students and housewives. Tables 6.1 and 6.2 represent two hypothetical cases, with rather different characteristics.

The numbers in each table represent the mean contribution for each condition, per type of subject. In Table 6.1, the factor "type of subject" *combines additively* with the factor "incentives" to determine the contribution levels. Comparing the results of different subpopulations (e.g., students vs. housewives) will provide us with an accurate estimate of how much of the contribution is due to which factor, but testing the effect of

Table 6.2. *Interaction*

	No	Low	High
Student	8.0	7.0	4.0
Housewife	2.0	3.0	6.0
Total average	5.0	5.0	5.0

the incentives by keeping the type of subject constant (i.e., testing the hypothesis that incentives matter in just one population of subjects, say, students) would not affect the validity of the main result – at least qualitatively. Randomizing across the "background" factors will not affect the main result either, as the aggregate means reflect the rankings as well as the differences in each population.

Consider now Table 6.2: the "type of subject" factor here *interacts* with the experimental treatment. Keeping that factor constant will provide information that is valid only for one particular type of subject. Any projection of that result outside that particular context will be unwarranted. Randomization will result in even more confusion. It is quite possible, as in my example, that the aggregate means do not reflect (neither quantitatively nor qualitatively) any of the specific populations' characteristics. In fact, experimenters usually perform some routine checks post hoc, just to make sure that certain variables (e.g., gender, age, education) do not affect the results significantly. However, this obviously begs the question posed by the "other factors" problem. The problem arises precisely because we may not have a good enough list of all the relevant factors, and post hoc checking is effective only if such a list is available.[22]

So, if randomization does not automatically or necessarily solve the "other factors" problem, do we have to surrender empiricism? Do we have to conclude that we can never know whether *e* supports *H*? Fortunately for us – and for science in general – we do not have to give up so easily. To see why, remember what I said repeatedly in previous chapters: empirical science cannot rely on deduction only. It must make use of a logic of inductive inference. Now, is *the mere possibility* of the existence of unknown factors or conditions that would invalidate the inference from *e* to *H* a *good reason* not to believe in the evidential claim? Clearly not. If it were a good reason, we would be endorsing a deductivist standard. We would be implicitly imposing the requirement, in other words, that the only good inference is a deductive inference – an inference such that if the

[22] It's important to stress, however, that the result of a randomized experiment is always of some interest (regardless of possible interactive effects) for it tells us, minimally, what happens on average in a certain population, defined by means of whatever criteria or causal concepts we happen to be using at present. This kind of information may be extremely valuable for practical purposes (e.g., when we have to decide whether to administer a drug to patients of a certain kind), even though we lack a deeper understanding of the causal mechanisms at work. I should thank John Dupré for pointing this out; see also Dupré (1984; 1993, Ch. 9). Hausman (unpublished) includes a useful discussion of the related problem of justifying practical causal generalizations.

premises are true, then the conclusion cannot possibly be false. In science, the mere possibility of a mistake is not a good enough reason to question an inference or claim. We make use of inductive, ampliative inferences, and such inferences always leave open the possibility of error. We need more than possibility and less than certainty: we need probability. And to assess the probability that a given experiment or inference from e to H is flawed, we need at the very least to state what sort of flaw or disturbing factor we are worried about. But once we have done that, we can try to control that factor, or correct the potential flaw experimentally by means of a new, improved design.

Another way to put it is in terms of the empirical support that an alternative hypothesis has to receive before it can be taken as a serious rival. The hypothesis "there is an unknown flaw in this experiment" cannot possibly receive any empirical support, according to the criterion of inductive inference defended in this chapter. In fact, such a hypothesis is compatible with any amount of evidence that can possibly be collected, regardless of whether the hypothesis is true or false. (Whatever evidence we collect, the probability of obtaining such evidence, given that the "unknown flaw" hypothesis is false, is always very high.[23]) Because an empiricist wants to make decisions based on good empirical evidence, the "unknown flaw" hypothesis can never constitute a good reason not to accept an experimental result.

The "other factors" problem, to sum up, is an insurmountable problem only for the deductivist – the philosopher or scientist who believes that deduction is the only legitimate method of scientific inference. Genuine deductivists are extremely rare, because the deductivist position is pragmatically impossible and does not stand up to rigorous scrutiny. Scientists make use of nondeductive inferences all the time, as we also do in everyday life. Still, we tend to make the mistake of modeling the logic of inductive inference on that of deduction, and we are easily trapped by arguments like the "other factors" objection.[24] Once we have discovered the "deductivist trick" that is being played on us, however, it is easy to realize that arguments of this sort should be resisted. Science is fallible: it is logically *possible* that we are mistaken. Our job is to make sure that it becomes unlikely.

---

[23] I must thank Deborah Mayo for suggesting this way of framing the argument.

[24] Attacks to inductive methods based on implicit deductivist intuitions or requirements are quite common. Harry Collins's (1985) "experimenter's regress" is a well-known case. See especially his "mature" position in Collins (1994, pp. 501–2).

Summary and conclusions

In the previous chapter, I outlined the problem of testing, or how to draw reliable inferences from empirical evidence. Because the evidence by itself usually does not indicate the hypothesis we are interested in, we must impose further conditions on inductive inferences from e to H. I first discuss the requirement of predictive success, a popular idea among scientists and philosophers alike, and then introduce the debate on the PR phenomenon as a test case. I show that predictive success does not seem to be adequate as a requirement on inductive inference, and in light of what I say in this chapter, we can now see why. Remember the Duhemian argument from coincidence: the inference from predictive success to the truth of a hypothesis is supported by the consideration that consistent predictive success is highly unlikely, if the hypothesis were false. We can generalize this type of inference to all inductive inferences as follows: an inference from e to H is strong if and only if it comes from a test situation or setup such that the observation of e would be probable if H were true, but unlikely if it were false. This requirement is present, in one form or another, in most theories of confirmation. Given that this is not a book on induction, I try to be as noncommittal as possible on the precise way in which the requirement should be specified, and about its role in the context of a general theory of inductive inference. What I say, however, should be enough to account for the fact that experimenters are usually busy unpacking $\sim H$, that is, trying to specify the different possible ways in which we may make a mistake in inferring from e to H. Each alternative account of the experimental evidence leads to a new experimental design aimed at controlling the background factor or factors that may lead us astray.

By analyzing the structure of the argument from predictive success, then, we can appreciate the key role played by background assumptions. It is such assumptions that raise the problem highlighted by Duhem and Quine, and a solution must recognize their function, as experimental scientists do. In this chapter, I discuss a prominent approach that does take the background seriously: subjective Bayesianism. Subjective Bayesians, however, leave too much freedom for scientists to choose their favorite prior beliefs about the background conditions. As a consequence, it is always possible for a dogmatic scientist to stick "rationally" to a certain belief in spite of the evidence (or it is possible for some time, to be more precise). I argue then, following some recent proposals in the philosophy of confirmation and testing, that we should start from an "objective" approach to the evidential relation. Whether a piece of evidence supports a given hypothesis or not is an objective matter, which depends on whether the experiment has been performed correctly rather than on what scientists think about the correctness of the experiment.

Of course, not all subjective elements are eliminated by such a move: in each given experiment we can never know for sure whether all the relevant sources of error have been controlled for. The model of the ideal controlled experiment, in which the inference from evidence to hypothesis is maximally strong, can, however, guide us as a normative ideal that we should try to approximate. Instead of simply following our arbitrary priors, the model prescribes to systematically check the potential sources of error recognized by the scientific knowledge of our time, hoping that other as yet unrecognized mistakes are detected during pilots. To illustrate, I discuss some experiments run in the last two decades in order to check some possible experimental errors in the "classic" PR experiments. The last two sections include some general reflections on the nature of the evidential relation and a reply to some common challenges.

PART TWO

INFERENCES FROM THE EXPERIMENT

SEVEN

External Validity

The first part of this book is devoted to inferences *within* the experiment. Among such inferences, I single out causal ones as especially important. The experimental method is the most powerful tool for finding out about causal relations. Experiments, in principle, allow the variation of a putative causal factor while keeping all the other relevant circumstances fixed, so as to observe the effect of that factor acting alone on the system under study. By iterating this procedure, the influence of all the putative causes can, in principle, be studied and various hypotheses tested. In the laboratory, such an investigation can be carried out in privileged conditions, under which background circumstances can be kept constant, disturbing factors shielded, and putative causes triggered at will.

Experimental scientists thus look for causal relations in very special settings, which are rarely if ever instantiated in the "real world" outside the lab. However, often they are not interested in what happens under these circumstances per se. Rather, they want to extrapolate from the specific experimental setup to learn something of more general applicability. When medical researchers investigate the effects of a drug on laboratory rats, they are usually looking for a result that can be generalized from mice to men, to cure fellow human beings suffering from some condition. Similarly, most economists are not particularly interested in what happens when a group of undergraduate students play lottery games. They would like to learn something about the functioning of markets, about economic behavior in the real world. However, how do they infer from the special circumstances created in the laboratory to the phenomena that take place "in the wild"? This question – the problem of *external validity* – is at the center of the second part of the book. In this chapter, I outline the problem in very general terms and discuss some arguments that can be found in the experimental literature. The status of these arguments is not always clear. Sometimes they downplay the relevance of external validity by suggesting that the application of simple methodological rules prevents the problem from arising in the first place. Other arguments tackle the problem more constructively, by specifying a series of conditions that make an experimental result valid outside the laboratory. Although none of these arguments is decisive, jointly they offer a number of extremely useful insights and can be used to construct an account of how experimental results can be generalized outside narrow laboratory conditions.

The natural and the artificial

To write on external validity is challenging. Philosophers of science, surprisingly, have very little to say about it. Experimental economists also tend to ignore or downplay the relevance of external validity; they typically say that it is not a particularly useful concept and, moreover, that worrying too much about it may turn attention away from more important issues of experimental design (cf. e.g., Plott 1987, 1999). Because there is nothing worse than inventing a pseudoproblem and then wasting time to solve it, I must first of all defend the relevance of external validity. And in order to do that, it is necessary to clarify the meaning of this concept.

The concept of external validity, as used in psychology and social science in general, is best defined by way of a contrast. *Internal* validity is achieved when some particular aspect of a laboratory system (a cause–effect relation, the way in which certain factors interact, or the phenomena they bring about) has been properly understood by the experimenter. For example: the result of an experiment E is internally valid if the experimenter attributes the production of an effect Y to a factor (or set of factors) X, and X really is a cause of Y in E. Furthermore, it is *externally* valid if X causes Y not only in E, but also in a set of other circumstances of interest $F, G, H, \ldots$[1] Problems of internal validity are usually chronologically

---

[1] The internal-external validity distinction is discussed primarily in social science books on research methods and goes back at least to Campbell and Stanley (1966); see also Cook and Campbell (1979). A finer set of categories is sometimes used to distinguish different dimensions of external validity (see e.g., Christensen 2001, Ch. 14): *population validity* (generalizing to a different population of subjects), *ecological validity* (generalizing to the behavior of the same subjects in different circumstances – the terminology is due to Brunsvik 1955) and *temporal validity* (generalizing to the same population, in the same circumstances, but at a different time). On external validity in psychology, see also Kruglanski (1975), Henshel (1980), Berkovitz and Donnerstein (1982). The same problem arises in different guises in other sciences; on biochemistry (in which it is known as the in vitro–in vivo problem), see for instance, Strand, Fjelland, and Flatmark (1996).

and epistemically antecedent to problems of external validity: it does not make much sense to ask whether a result is valid outside the experimental circumstances unless we are confident that it does therein.

To appreciate the relation between internal and external validity, consider that typically a laboratory experiment solves problems of *scale* and *variation*. Many spontaneous phenomena are just too big or too small to be investigated in their natural settings. An epidemic, for instance, may be too large and last too long to be studied effectively in the field, but in the laboratory, we can reproduce its key features in a population of animals (guinea pigs, mice, flies) that are relatively inexpensive, whose life cycles are shorter, and which can be stacked (and killed) in great numbers inside laboratory cages.[2] Experimental studies of evolution follow a similar strategy: fruit flies reproduce so rapidly that hundreds of generations can be examined in just a few months' time, whereas a similar study of, say, human inheritability would take thousands of years. What is smaller often is also simpler: scientists working on the genome project mapped the DNA of the bacteria *Escherichia coli* first because it is much shorter than the human genome. At the other end of the spectrum (where real-world phenomena are too small), there are things like electrons and viruses, too small and elusive to be detected in uncontrolled circumstances but more easily tamed in laboratory conditions with the aid of powerful instruments of observation.

The second problem, variation, has to do more directly with causation. As noticed in Chapter 4, it is difficult to make causal inferences when there is either too much or too little variation. To find out what causes what, you need just the right amount of variation: only one factor has to vary while the others remain fixed at some specific level. Outside laboratory conditions, this happens quite rarely – in fact, it happens in so-called natural experiments only. Of course, we often draw causal inferences from nonlaboratory evidence by *pretending* that certain factors did not change, or by subtracting the effect of one confounding variation from the data and then trying to figure out "what would have happened if" only a putative cause had varied. But such procedures are usually more risky and provide less reliable results than genuine controlled experiments.

To clarify the relation between external validity and the problem of causal discovery, recall the principle of probabilistic causation introduced in Chapter 4: a necessary condition for X to be a cause of Y is that X raises the probability of Y at least in *some* causally homogenous

---

[2] On the "laboratory revolution" in medicine, see the studies in Cunningham and Williams (1992).

background conditions. To say that X causes Y in E, then, is to say that X raises the probability of Y in *that* particular context – when the other factors are kept not only constant but fixed *at that particular level*. (Ingesting an aspirin raises the probability of reducing headache only if the relevant circumstances are "right.") Because causal relations are context specific, the result of an experiment should always be expressed as a ceteris paribus claim: A causes B, other things being "right." For methodological and epistemic reasons, a full list of the "right" conditions can rarely be specified: experimenters try to simplify the experiment by omitting certain factors and conditions that may be at work in the real world and by "shielding" the experimental system from random disturbances. Moreover, many unknown, potentially disturbing factors are dealt with by randomizing across the control and experimental groups. What happens when things are not "right," however – and exactly what "right" means in the first place – is usually left open and has to be investigated separately.

For these reasons, there exists a trade-off between the two dimensions of experimental validity. The stronger an experimental design is with respect to one validity issue, the weaker it is likely to be with respect to the other. The more artificial the environment, the better for internal validity; the less artificial, the better for external purposes. Notice that the problem is not just that, say, mice are different from men or students in economics may be different from the managers of a big multinational firm. There are at least two other important disanalogies between experimental and real-world conditions. First of all, *laboratory* mice are very different from ordinary mice. Medical researchers work on "animal models" that have been constructed especially for laboratory purposes. The so-called standard fruit fly (*Drosophila melanogaster*), to name a well-known example, is not the insect flying around the bananas in your kitchen. It is a carefully selected organism that does not exist in nature and is artificially stabilized in spite of its tendency to evolve into something else (see Kohler 1994). Similar remarks apply to chemistry, in which the "preparation of materials" is a key ingredient of experimental practice. In experimental social science, the "preparation" of subjects is limited by obvious ethical constraints. However, it does take place, sometimes by selecting those individuals who display certain characteristics (e.g., a particular attitude toward risk – cf. Roth and Malouf 1979) or by eliminating those who display ambiguous traits or "confused" behavior. Sometimes subjects are "prepared" in the sense that they receive very precise instructions or cues about how they *should* behave in the experimental conditions.

Secondly, laboratory entities act and live (and often die) in tightly controlled conditions, which are very different from those found in nature and in our society. In the social sciences, this difference is particularly evident if one looks at the abstract tasks that subjects are customarily asked to perform in the laboratory.[3] These various dimensions of the laboratory-versus–real-world gap are worth keeping in mind. Both critics and apologists of experimentation in the social sciences often focus on the statistical representativeness of the experimental population as if it were the most relevant problem to be solved. It is, in fact, just one aspect of the external validity issue – which is in reality much more complicated than that.

Metaphysics

External validity is surprisingly absent from philosophical discussions of experiment. This is partly because philosophers, with few exceptions, take physics as the paradigmatic experimental science, and physicists tend not to recognize external validity as a separate inferential problem on its own. There are historical, sociological, and metaphysical reasons for this. I won't say much on the historical and sociological ones, except that physics has gone through at least two "laboratory revolutions": from an Aristotelian science concerned with the explanation of spontaneously occurring phenomena by means of unaided observation, to a Galilean science investigating natural phenomena in the "ideal" conditions of the laboratory, to a science, finally, whose main questions and answers stem from laboratory work. Most research in contemporary experimental physics begins in the lab, is carried on in the lab, and ends in the lab. Every now and then, of course, some stunning application is developed that "exports" an experimental result to other domains. However, these are occasional (albeit extremely important) side effects of what is otherwise "pure" laboratory science (cf. Hacking 1992, p. 33).

The generalizability of pure laboratory science is sometimes simply assumed by means of metaphysical speculation. Isaac Newton, for instance, is often quoted for his view that "the qualities [. . .] which are found to belong to all bodies within the reach of our experiments, are to be

---

[3] The environment itself is a complex entity, whose characteristics can be grouped in various ways. Glenn Harrison and John List (in press), for example, classify possible discrepancies between the laboratory and the field along six dimensions: subject pool, information, commodity exchanged in the experiment, nature of the task, nature of the stakes, and nature of the environment.

esteemed the universal qualities of all bodies whatsoever" (1687, p. 398).
And the great physiologist Claude Bernard was probably influenced by
Newton when he wrote that "all animals may be used for physiologi-
cal investigations, because with the same properties and lesions in life
and disease, the same result everywhere recurs" (1865, p. 115). To make
sense of these claims, we can imagine Newton and Bernard appealing to a
very basic metaphysical principle: should exactly the same circumstances
repeat twice, the same effects will follow from them (*same cause, same
effect*).

This *principle of the uniformity of nature*, however, does not dissolve
even the most common external validity worries. The psychologist Baruch
Fischhoff uses a graphic example to represent the real terms of the problem of external validity; in the lab, choices look like this:

Choice A In this task, you will be asked to choose between a certain loss and a
gamble that exposes you to some chance of loss. Specifically, you must choose
either: Situation A. One chance in 4 to lose $200 (and 3 chances in 4 to lose
nothing). OR Situation B. A certain loss of $50. Of course, you'd probably prefer
not to be in either of these situations, but, if forced to either play the gamble (A)
or accept the certain loss (B), which would you prefer to do? (Fischhoff 1996,
p. 232)

But in the real world, choices look like *this*:

Choice B My cousins . . . ordinarily, I'm like really close with my cousins and
everything. My cousin was having this big graduation party, but my friend – she
used to live here and we went to . . . like started preschool together, you know.
And then in 7th grade her stepdad got a job in Ohio, so she had to move there.
So she was in Ohio and she invited me up for a weekend. And I've always had
so much fun when I'd go up there for a weekend. But, it was like my cousin's
graduation party was then, too – like on the same weekend. And I was just like
I wanted to go to like both things so bad, you know. I think I wanted to go more
to like up Ohio, you know, to have this great time and everything, but I knew my
cousin – I mean, it would be kind of rude to say, "Well, my friend invited me up,
you know for the weekend." And my cousins from out of town were coming in
and everything. So I didn't know what to do. And I wanted mom to say, "Well,
you have to stay home," so then I wouldn't have to make the decision. But she
said "I'm not going to tell you, you have to stay home. You decide what to do."
And I hate when she does that because it's just so much easier if she just tells you
what you have to do. So I decided to stay home basically because I would feel
really stupid and rude telling my cousin, well, I'm not going to be there. And I
did have a really good time at her graduation party, but I was kind of thinking I
could be in Ohio right now. (ibid., p. 232)

What do choices in environments like the former tell us about behavior in situations like the latter? One possible answer is that experiments like Choice A test subjects' "pure" decision-making capacities. However, this is still unsatisfactory: the decision processes may be completely different in the two circumstances, and "purity" is a poor consolation if it is unlike anything that we are ultimately interested in explaining and understanding. A more appealing answer is that in situations like Choice B, there is just too much going on, and simplified settings like Choice A are intermediary steps on the way toward the understanding of complicated real-world decision making. However, then the problem takes a completely different form, one that defies simple metaphysical postulation. Charles Plott formulates the issue appropriately as follows:

What use are experimental results to someone who is interested in something *vastly larger and more complicated*, perhaps *fundamentally different* than anything that can be studied in a laboratory setting? (1987, p. 193, my emphasis)

"Same cause, same effect" won't do because the conditions are rarely, if ever, the same.

Eliminating false theories

We'll come back to the uniformity of nature principle later on. In the meantime, it is necessary to review a set of arguments that tackle external validity from a different (*methodological*, rather than metaphysical) angle. Consider, for instance, the following claims by Graham Loomes and Louis Wilde, respectively:

If one or more of the fundamental axioms of expected utility theory fail under such [i.e., experimental] apparently favourable conditions, there are surely grounds for questioning the power of the model as a *general* theory of individual decision-making under risk and uncertainty. If the basic axioms are substantially and *systematically* violated in these simple cases, how confident can we be about their validity in more complex cases? (Loomes 1989, p. 173)

If an experiment includes all parameters relevant to a particular theory, and if the theory fails to predict well in the simplified setting of the laboratory, then it cannot be expected to predict well in more complex environments. [. . .] The experiment does not need to be "realistic" and no presumptions need be made about its connection to more complex ("real-world") environments [in order to use laboratory experiments to reject some theories as nonsense]. (Wilde 1981, p. 143)

John Hey seems to follow a similar line of reasoning when he represents laboratory experimentation as a way of performing a preliminary selection of theories ultimately intended to explain real-world phenomena. According to Hey, experimental economics allows the separate treatment of two crucial but distinct issues, namely

1. that the theory is correct given the appropriate specification (that is, under the given conditions); 2. that the theory survives the transition from the world of the theory to the real world. (Hey 1991, p. 10)

A theory failing at stage one, according to Hey, should not be allowed to enter stage two of testing. Such "falsificationist" arguments convey an important idea: that external validity worries are not (and should not) be raised for *every* economic experiment. Real-world applicability can be the goal of a whole *research program*, and it is perfectly reasonable to focus on the investigation of relatively simplified or abstract settings in the early stages of the program itself. Plott also endorses this approach to experimentation:

The logic is as follows. General theories must apply to simple special cases. The laboratory technology can be used to create simple (but real) economies. These simple economies can then be used to test and evaluate the predictive capability of the general theories when they are applied to the special cases. In this way, a joining of the general theories with data is accomplished. (Plott 1991, p. 902)

A staggeringly large number of theories exist. One purpose of the laboratory is to reduce the number by determining which do not work in the simple cases. The purpose is also to improve the models by exploring how a model might be changed to make it work better in the simple cases. General models, such as those applied to the very complicated economies found in the wild, must apply to simple special cases. Models that do not apply to the simple special cases are not general and thus cannot be viewed as such. (ibid., p. 905)

These pronouncements seem to presuppose a fairly radical form of falsificationism. Let us grant, for the sake of argument, that laboratory testing may work as a *screening device* to reject the theories that fail to describe what goes on in simple laboratory circumstances. A strict falsificationist attitude unfortunately does not help whenever we want to infer from a *positive* experimental finding (that a certain theory, model, or hypothesis works in the laboratory) to nonexperimental circumstances. If real-world applicability is the ultimate goal in (laboratory and nonlaboratory) science, then this is no minor flaw.

It is, of course, just a specific instance of a flaw already highlighted in Chapter 3: falsificationism is too thin a methodology for science. *Pace* Popper, scientists are not merely concerned with the refutation of theories, they also want to learn something positive about the applicability of their theories and the truth of their hypotheses. Scientific method must include an inductive element as well as a deductive one, but then what does the success of a theory or hypothesis in the laboratory teach us about its success in other nonexperimental circumstances? The arguments above tell us that we should get things right in the lab first and that this is a necessary condition for getting things right in the outside world, but certainly they cannot (and in fact do not) claim that getting things right in the lab is a *sufficient* condition for the generalization of results to other circumstances.

Notice, however, that even a *negative* response during the early stages of a research program does not by itself indicate that a theory is *completely* on the wrong track. It is perfectly conceivable, in fact, that a model that has been "refuted" under laboratory conditions may nevertheless work under other (real-world) circumstances. The experiment could lack some factor or condition that is crucial for the model's applicability, a factor that is instead present under other (nonlaboratory) circumstances. If this seems a bit speculative, consider the case of preference reversals again: in the 1980s and 1990s, several experimenters conjectured that mechanisms like repetition or arbitrage, which were absent from the first generation of experiments, could eliminate or reduce the rate of reversals and make the theory of expected utility applicable "in the wild" despite its laboratory failures (e.g., Berg, Dickhaut, and O'Brien 1985; Chu and Chu 1990). Such attempts were clearly aimed at saving expected utility theory from refutation, in apparent contradiction to the falsificationism of Loomes, Wilde, Hey, and Plott's remarks.

These arguments, therefore, must rely on an implicit but fundamental assumption: a theory that lacks a *complete* specification of the conditions or factors that make it applicable is somehow unsatisfactory and should be replaced. This is probably what Loomes, Wilde, and Plott have in mind when they stress that scientific theories should be general or universal in scope of application.[4]

---

[4] This terminology might be a little confusing for philosophers of science, who tend to distinguish sharply (albeit perhaps pedantically) between universality and generality requirements. Universality is commonly understood as a purely syntactic feature ("For *all* objects of type X, ..."), whereas generality is a matter of scope of application and depends especially on the sort of entities and properties cited in a given theory or scientific law (a theory of mammals clearly has wider scope than a theory of horses, a theory of electromagnetism has wider scope than a theory of electricity, and so on). Although in this chapter I focus exclusively on what philosophers call "universality," it is worth keeping the distinction in mind because lack of universality and lack of generality are limitations of entirely different kinds.

The idea that the search for universal scientific theories (or laws) drives scientific research is sophisticated and requires an articulate discussion. It is also dangerous material for economic experimenters, for it cuts both ways and might in the end turn against experimental economics itself. The same concepts used to defend experimental economics (the goal of universality and the method of falsification) are, in fact, also invoked occasionally to *deny* the fruitfulness of the experimental approach. In May 1999, for example, an aggressive critique of the very idea of laboratory experimentation was published in *The Economist*. The article concluded as follows:

Whether experiments improve the scientific credentials of the discipline must be very much in doubt. In the end, the main problem is not that designing good experiments is hard. It is simply that, unlike physics, economics yields no natural laws or universal constants. That is what makes decisive falsification in economics so difficult. And that is why, with or without experiments, economics is not and never can be a proper science. (Economics Focus 1999, p. 96)

But why should a scientific statement (a law, a theory, a causal claim) be universal in character?

Universality, sufficiency, completeness

Universality is a purely syntactic feature of scientific laws or theories, but with very interesting methodological implications. Its main virtue is that *a universal conditional statement carries its domain of application written in its antecedent*. Consider the simple example "For all x, if x is a swan and is of European breed, then x is white." Such a claim can be appropriately tested *only* in the domain of European swans. The observation of an Australian black swan would not falsify it, in other words, for it would fall outside its domain of application. To impose the requirement that scientific theories should be universal, therefore, is equivalent to imposing the requirement that the antecedent of their laws should specify a set of conditions that are jointly *sufficient* for the instantiation of the effect. Another way to put it is that good theories should be *complete*, or should carry their domain of application written in their assumptions.

From such a standpoint, it follows quite naturally that an experiment reproducing *all* the initial conditions or relevant assumptions of an economic model must be a "good" experiment by default. If the model does not apply in those experimental conditions, we want to know *why*. Vernon Smith puts it as follows:

If [an experiment's] purpose is to test a theory, then it is legitimate to ask whether the elements of alleged "unrealism" in the experiment are parameters in the theory. If they are not parameters of the theory, then the criticism of "unrealism" applies equally to the theory and the experiment. If there are field data to support the criticism, then of course it is important to [modify] the theory to include the phenomena in question, and this will affect the design of the relevant experiments. (Smith 1982, p. 268)

According to Plott, experimental economics' revolutionary achievement consisted of shifting the focus from whether a certain experiment reproduces a real-world system accurately to whether it accurately tests a theory or model. Experimental economists have thus removed two "constraints" that stood in the way of laboratory research:

The first was a belief that the only relevant economies to study are those in the wild. The belief suggested that the only effective way to create an experiment would be to mirror in every detail, to simulate, so to speak, some ongoing natural process. [. . .] As a result the experiments tended to be dismissed either because as simulations the experiments were incomplete or because as experiments they were so complicated that tests of models were unconvincing. [. . .] Once models, as opposed to economies, became the focus of research the simplicity of an experiment and perhaps even the absence of features of more complicated economies became an asset. The experiment should be judged by the lessons it teaches about the theory and not by its similarity with what nature might have happened to have created. (Plott 1991, p. 906)

This argument has the undeniable rhetorical advantage of sending the ball back into the theorist's camp, so to speak. However, we must be careful not to misinterpret what Smith and Plott are saying. In this book, the term *theory* is used to refer only to the precisely stated, coherently organized, possibly formalized parts of economic knowledge, which is also how economists use the term when they want to be precise; as already mentioned, you don't get an informal speculation published in the *Journal of Economic Theory*. An informal hypothesis, a rough conjecture, or a guess is not a theory (although it can perhaps be turned into a theory and can often be tested just like a prediction derived from a formal model). However, under this interpretation of *theory*, the requirement that theories be modified in order to incorporate every factor that may affect their applicability is highly unrealistic.

Let us take a simple example. We know that in certain experimental conditions – such as the familiar case of preference reversals – the transitivity principle of utility theory seems to break down. Taking the universality requirement seriously implies that the principle should be amended so as to make sure that the conditions in its antecedent ("for all human beings, if X then Y") are truly sufficient for the consequent to be instantiated. In fact some economists suggest that factors such as learning (by repetition of the experimental task) and arbitrage can restore the transitivity of preferences (e.g. Berg et al. 1985, Chu and Chu 1990). But there is as yet no generally accepted theory of how to model the effect of repetition and arbitrage on the transitivity of preferences. According to an extreme reading of Smith and Plott's remarks, theorists should invest more energy in the task of formally incorporating such mechanisms into economic theory, for the domain of a theory must be written in the antecedent of its laws. Until such mechanisms have been formally modeled, experimenters are entitled to consider the theory tested and refuted.

However, this interpretation is clearly misguided. To begin with, it is not difficult to think of ways of modelling the effects of repetition and learning; yet, economists do not seem to give it much importance, which suggests that the incompleteness of economic theory is relatively unproblematic for them.[5] Second, we must remember that theory testing is only *one* possible function of experiments. Many experiments are devoted to investigating the robustness of economic phenomena, and this activity is only partly guided by theory. Economists are aware of the limitations of theoretical modeling and make sure that such limitations do not become obstacles to empirical research. The fact that utility theory is not expanded to include the effect of arbitrage and repetition, for example, should not be an embarrassment to economists. Take a physical law like $F = G(m_1 m_2 / r^2)$. The law does *not* state that "if two bodies have masses m_1 and m_2, and lie at a distance r, then the force of attraction between them is directly proportional to the product of their masses and inversely proportional to the square of the distance." It states that "if two bodies have masses m_1 and m_2, lie at a distance r, *and no other force than gravitation intervenes*, then ... (etc.)."

The standard way of dealing with such exceptions is to include a ceteris paribus (cp) clause in the antecedent of laws: $(P\&cp) \rightarrow Q$ – "if P then Q, other things being equal."[6] However, notice that to a certain extent, the

---

[5] I owe this point to Bob Sugden (in correspondence).

[6] The best discussion of ceteris paribus clauses in economics to date can be found in Hausman (1992a, Ch. 8); for more recent surveys, see Mäki and Piimies (1998) and Boumans and Morgan (2000). The need for a ceteris paribus interpretation of even the most general physical theories has been forcefully argued by Nancy Cartwright (1983). It is also worth mentioning that theoretical simplifications are not the only simplifications involved in laboratory science: experimenters often *simplify theories* in order to derive testable predictions, and in this sense, experiments are twice removed from the reality that economists try to model. I must thank Joep Sonnemans for pointing this out.

domain of the theory is left vaguely specified. The ceteris paribus clause is a "catchall" proviso; although certain factors in its domain may be known (electromagnetic effects for the attraction between bodies, arbitrage and repetition for the transitivity principle, etc.), others are willingly left unspecified.

A staunch supporter of the universality requirement, of course, does not have to give up so easily. The classic rejoinder is that "laws" such as "ceteris paribus, for all bodies $F = G(m_1 m_2 / r^2)$" are unsatisfactory, or that the ceteris paribus clause merely points to a problem that ought to be solved by means of a better theory. One should "unpack" the clause, in other words, and include explicitly all the relevant factors in the assumptions of the theory; that is, one must move from $(P \& cp) \rightarrow Q$ to $(P_1 \& P_2 \& P_3 \& \dots P_n) \rightarrow Q$.

However, the requirement of modeling *all* relevant causal factors in the antecedent of theories/laws does not seem to be very promising for sciences like economics. Economic concepts refer to entities and phenomena that are evidently nonfundamental – that *supervene* on other entities and phenomena that are customarily studied by other disciplines. The conditions for the law of demand to hold, for instance, include the existence of human beings with certain preferences/beliefs but also with certain cerebral functions, which in turn depend on certain chemical laws, which depend on certain physical laws. It seems unlikely that we can model all this stuff in an economic theory because (1) we have only a vague idea of how the different levels (physical, chemical, biological, psychological) are related to one another; (2) in order to model all the relevant conditions for the applicability of the law of demand, we should go far beyond the linguistic resources of standard economic theory; (3) if we knew how to do that, the resulting theory would probably turn out to be terribly complicated; and finally, (4) the theory would not be of much use, because eventually we want variables that we can control for policy purposes, and these usually lie at the level of analysis of standard economic theory.

In an often-quoted passage, Jerry Fodor takes such features to be the distinguishing characteristics of the "special sciences" (as opposed to fundamental sciences like, say, the physics of small particles):

Exceptions to the generalisations of a special science are typically inexplicable from the point of view of (that is, in the vocabulary of) that science. That's one of the things that makes it a special science. (Fodor 1987, p. 6)

Special science laws are unstrict not just de facto, but in principle. Specifically, they are characteristically "*heteronomic*": you can't convert them into strict laws by elaborating their antecedents. One reason why this is so is that special science laws typically fail in limiting conditions, or in conditions where the idealisations presupposed by the science aren't approximated; and, generally speaking, you have to go outside the vocabulary of the science to say what these conditions are. (Fodor 1989, p. 78)[7]

To pursue a complete enumeration of the conditions covered by a ceteris paribus clause would force scientists to move beyond their favorite realm of phenomena and level of analysis. The neoclassical economist, for example, would have to abandon her models of rational economic agents and engage in a much deeper analysis of human psychology. The laws of psychology being incomplete in character, one would have to move one step further down the ladder of microfoundations, to, for example, neurophysiology. Eventually, according to the ideal reductionistic picture, all the laws of the special science would find their justification in the generalizations of physics.[8]

Of course, the feasibility of a reduction of economics to psychology and of the latter to physics cannot be ruled out in principle, but some empirical evidence must be presented before we accept reductionism as a fundamental ingredient of our methodology. (We cannot gamble the whole scientific method on a mere *possibility*.) Special sciences such as economics exist precisely because it would not be practical to try to explain right from the start economic phenomena in terms of more fundamental laws, such as those of psychology or even physics (if these laws exist at all, of course). However, if the criterion for scientificity is sufficiency or completion, then "the only real science is basic physics" (Fodor 1987, p. 5).[9]

---

[7] Cf. also Fodor (1974).

[8] See also Kincaid (1996, Ch. 3) for a reductio ad absurdum of this sort, with particular reference to the social sciences.

[9] If such arguments do not sound immediately convincing, it partly has to do with the fact that we are so used to thinking in reductionistic terms that we tend to ignore or underestimate the practical difficulties with which even the most basic reductions are achieved. Contrary to common scientific propaganda, for example, even allegedly unproblematic reductions, such as those from chemistry to physics or from biochemistry to chemistry, are incredibly messy and often just not feasible. Consider, moreover, that science in its historical development does *not* display a progressive reduction in the number and variety of established theories. On the contrary, new disciplines and subdisciplines are constantly created, and phenomena are explained by means of an increasing number of theories and models at different levels of specification. Science becomes increasingly varied and specialized, instead unified under more and more fundamental theories. For the argument on the increasing variety and "disunity" of science, cf. Suppes (1984). On the limitations of "horizontal" reductionism, see also Dupré (1993) and Cartwright (1999). Mäki (2001b) offers a useful discussion of the unificationist project in economics.

Of course, *some* important steps in the progress of science have been achieved by enlarging the domain of a theory and digging "deeper" into the microstructure of reality. Classic examples are the unified explanation of light and magnetism by means of electromagnetic theory and the explanation of heat and pressure by means of statistical mechanics. As far as experimental economics is concerned, some promise seems to lie in the partnership with evolutionary theory and neurobiology. However, it is important to realize exactly what these examples can and cannot prove. As a critique of the critique of the sufficiency or universality requirement, these examples are misplaced: no one means to rule out or criticize a priori programs such as evolutionary economics and neuroeconomics. These programs may or may not be successful, and we are presently in no position to make serious predictions about this.[10] The point is that their success should not be made a *precondition* for doing "proper" economic science. Experimental economists *are* doing proper economics right now, and can keep doing economics without incorporating all the conditions that specify a theory's application within the theory itself. In fact, most science goes on like that, the integration with "lower-level" or neighbor theories being typically only partial and rough.

Universality requirements are motivated by a basic misconception of the role of theories and models in science. Although a more precise characterization will have to wait until Chapter 9, a few brief remarks will suffice for the present discussion. Theories are *tools* that help us overcome our limited cognitive capacities. Ronald Giere (2002) has recently proposed an illuminating analogy: theoretical models are like cognitive scaffoldings supporting our representations and inferences. As such, they cannot be too complicated: a model is usually a simplified and artificially isolated system constructed for analytical purposes. A ceteris paribus clause must be attached to any claim derived from analyzing a model because the model abstracts from the influence of several real-world factors by simply ignoring them (which is to be expected: if a model included all the complex features of a real-world system, there would be little point in studying the model in the first place). However, model-based knowledge must be applied intelligently, making all the adjustments that are required from case to case. The possession of the informal and practical knowledge required to put a theory to work is exactly what distinguishes a good scientist from the layman who just happens to read a scientific

---

[10] For a survey of the main results achieved in neuroeconomics so far, cf. Camerer, Loewenstein, and Prelec (in press), and Smith (in press).

textbook. The layman thinks that he or she can infer the domain of applicability from the theory itself but, in fact, has only a very vague clue of how to use it in concrete cases. A lot of training and experience are required in order to apply scientific theories.[11]

How are models applied, then, if they do not have the internal resources to specify their own domain of application? A popular view is that models must be put in correspondence with the real world by means of a *theoretical hypothesis* stating what kind of relation holds between a given model (or set of models) and a given real-world system (or set of systems).[12] Different philosophers have different views about the nature of this relation (similarity, approximate similarity, isomorphism, etc.), and I have nothing new or deep to say about it here. I'm even tempted to claim that perhaps different models are applied differently (they relate to their target systems in different ways), and hence there may be no general story to be told.

However that may be, the important point here is that the theoretical hypothesis is not part of the theory's axiomatics – the intended domain of a theory is defined by a separate component. Thus, for example, by learning game theory models, you cannot figure out exactly what their intended applications are. What you are given is a handful of standard textbook applications as examples of the sort of things the models are likely to be applicable to. The exact domain of application is not defined in the textbook because it is part of the informal theoretical hypotheses that are tested in applied science all the time. In "normal science," students learn some modeling techniques and are told that certain models can be applied to certain paradigmatic cases.[13] However, it is implicitly understood that the intended domain may vary, even though the models remain the same. This leaves the issue of the domain of scientific theories partly open – which is just as it should be, in my view.

Empiricism

We need another reading of Smith and Plott's position, then. Although the "strong" version of the universality requirement imposes unrealistic

---

[11] For an excellent example of the vast amount of non–theoretical knowledge required to apply scientific theories, see for instance the excellent recent books on auction theory by Paul Klemperer (2004) and Paul Milgrom (2004). Auctions will figure prominently among the examples of the next two chapters.

[12] See, for instance, Giere (1979, 1988).

[13] This way of interpreting the role of textbook science owes a lot to Kuhn's (1962) seminal work. See also the already cited paper by Cubitt (in press) for an excellent discussion of the problem of identifying a theory's domain of application and the related issue of the interpretation of laboratory tests of economic theories.

demands on scientific theories, there is a more reasonable, "weaker" reading of it that seems promising for our purposes. The idea, in a nutshell, is that even though we may be unable to model all the factors that limit the applicability of a theory, such factors must nevertheless be defined in a precise enough way to make them *amenable to empirical testing*. The rationale for this weaker requirement is well illustrated by Chris Starmer:

While potentially valid criticisms of a given experimental design, [the] objections [. . .] which point to a specific limitation of the experimental setting do not seem to tell against experimentation *per se* since the experimenter can mount a ready response to each such objection. For example, if the hypothesis is that "the free rider theory failed [in a given public goods experiment] because the incentives were too small," then run a new experiment with bigger incentives. If it is suspected that communication between subjects enabled them to "beat" the free rider problem, design a new experiment that makes communication more difficult. So long as the theory defender identifies some specific aspect of the design that renders it unsatisfactory as a test of the target hypothesis, it seems reasonable to think that a new experiment could be run that could "correct" the limitation of the earlier test. Hence, criticisms that point to specific reasons as to why an experiment is not a satisfactory test of a hypothesis do not tend to undermine experimenting; they suggest new problems that can be investigated experimentally; they enrich the experimental agenda. (Starmer 1999, p. 9)

Consider the analogy between this proposal and the way in which internal validity worries should be tackled. In Chapter 6, I argued that generic skepticism about the internal validity of a result is not to be taken seriously. One must point out *why* the result may not be valid and by doing so, implicitly provide a new hypothesis that is amenable to empirical testing. Similarly, the mere *possibility* that a result may lack external validity is irrelevant: we need probability in science. And one cannot even start to figure out what the relevant probabilities are unless some reason for the lack of validity is indicated. The hypothesis that a result might lack external validity because the experiment lacks a crucial (but unknown) feature of the real world cannot even be tested severely. Such a hypothesis can fit any amount of evidence that can ever be collected, even if it's false. Hence, it fails to respect the requirement that we have imposed on inductive inference, namely that the probability of observing confirming evidence be low, if the hypothesis is false.

The moral, then, is that it is necessary to investigate *empirically* which factors among those that may be causally relevant for the result are likely to be instantiated in the real world but are absent from the experiment (or vice versa). That an experimental and a real-world system should differ in some respects is just inevitable and rather uninteresting by itself; the interesting question is whether the differences are causally relevant.

The distinction between (1) requesting the formulation of an empirically testable explanation of the alleged lack of validity of a given economic result and (2) the stronger requirement that such explanation be incorporated in economic theory can also be used to cast the shadow of instrumentalism away. That a model is applicable *here* but not *there* is puzzling and needs to be accounted for in some way; one does *not* face a dilemma between a complacent instrumentalist attitude toward the truth of economic theories (à la Friedman 1953) and the unreasonable request that such explanation be formally modeled in the theory itself (see also Starmer 1999, p. 18, on this point).

This approach to the external validity problem helps to shed some light on one of the most obscure and controversial methodological statements in the experimental economics literature. Vernon Smith (1982) defines a key "methodological precept" of experimental economics[14] in a way that reminds one of the metaphysical principle (same cause, same effect) discussed earlier:

Parallelism: Propositions about the behavior of individuals and the performance of institutions that have been tested in laboratory microeconomies apply also to nonlaboratory microeconomies where similar *ceteris paribus* conditions hold. (Smith 1982, p. 936)

Parallelism is the precept of external validity, and strikes one as a fairly strange proposition. The precept defines some minimal conditions of validity that, admittedly, are usually not satisfied by experimental results: the real-world economies we are ultimately interested in are invariably different (much more complicated) than anything we can implement in the laboratory. However, the precept, following Friedman and Sunder (1994, p. 16), should be read as setting the *guidelines* for a constructive methodological approach to external validity problems. If we all agree on the minimalist metaphysical principle, the experimenter says, than any *specific* external validity challenge must indicate how and where the experiment differs substantially from the real-world system we are interested in investigating. When this has been done, the actual relevance of these differences can be checked by empirical means.

Eventually, in Smith's words, "*Which kinds of behavior exhibit parallelism and which do not can only be determined empirically by comparison studies*" (1982, p. 936, emphasis in the original). External validity is a conjecture that has to be established empirically in each case. The challenge,

---

[14] The other precepts (their contents and status) are discussed in Chapter 11.

then, is to define more precisely the sort of methods that can be used in order to draw reliable external validity inferences. Here is a statement of the program:

When the theory performs well [in the lab] you [. . .] think, "Are there parallel results in naturally occurring field data?" You look for coherence across different data sets because theories are not specific to particular data sources. Such extensions are important because theories often make specific assumptions about information and institutions which can be controlled in the laboratory, but which may not accurately represent field data generating situations. Testing theories on the domain of their assumptions is sterile unless it is part of a research program concerned with extending the domain of application of theory to field environments. (Smith 1989, p. 152)

External validity arguments work by combining experimental and field data. But how exactly does this partnership work? What makes a good "mix" between experimental and nonexperimental evidence? Answering these questions is the task of the next three chapters. In an old article, Rice and Smith (1964) try to use the Bayesian approach that we find so inadequate in the first half of the book. Because this approach has not gained many followers, I relegate a summary and critique of the Bayesian "solution" to external validity in Appendix B. However, this leaves us with practically nothing to build upon for a general account of external validity inferences. The project will have to move inductively, from concrete examples of inferences from the experiment, to a general account that rationalizes their essential features.

Concluding remarks

As a young discipline, experimental economics struggled to survive within a generally skeptical and dismissive scientific community. Downplaying difficult methodological problems, therefore, may have been a simple survival strategy. But the time is now ripe to be bolder and discuss openly the challenges facing the experimental approach. External validity is one such challenge.

Most economists aim toward understanding the functioning of markets – and markets are, with few exceptions, naturally evolved as opposed to artificially created entities. In order to study such entities, they use a number of different techniques, from theoretical modeling to econometrics, surveys, simulations, and last but not least, experiments. The use of experiments to study nonlaboratory entities raises the problem of external validity. External validity is not a peculiar problem of social science experiments – quite the contrary – but other scientists rarely remind us of how artificial their experimental settings are. We are constantly bombarded by newspaper announcements that some stunning discovery has proved a link between, say, the gene of a mouse and a dangerous disease. However, we are rarely told how far away we are from a valid explanation of that disease in human beings. Every single month, a new substance promising to cure cancer is announced in scientific journals. However, as we know all too well, sadly most of these "cures" do not survive further scrutiny and never materialize in our hospitals.

Sometimes external validity takes the form of "the in vitro–in vivo problem" (biochemistry), sometimes it is called "ecological validity" (psychology), and sometimes it is called "parallelism" (economics), but the issue is always the same. How are experimental phenomena "exported" from the laboratory to the outside world? Are we entitled to infer from such peculiar circumstances to what happens in other circumstances of interest? In this chapter, I argue that the problem is genuine and important and I review some attempts to solve it, or simply make it go away. Following Chris Starmer, Vernon Smith, and generally the empiricist approach of this book, I argue that the external validity problem is empirical in character and must be solved by appropriately combining field and laboratory evidence. In the next two chapters, I outline a more detailed account, modeled on economists' concrete attempts to draw convincing external validity inferences.

EIGHT

Economic Engineering

How should external validity problems be tackled? Because the philosophical and economic literature provides little help, I look for inspiration somewhere else: in the actual techniques used by experimental
economists when they engage in applied science. What do economists do
when they have to make sure that their laboratory results will work in the
field?

This chapter is primarily descriptive in character. I focus on a cele-
brated application of game theory and experimental economics to solve
a real-world problem: the construction in 1993–94 of a new market institution, by means of which the Federal Communications Commission (an
agency of the U.S. government, FCC from now on) allocated a peculiar
kind of good (telecommunication licenses). One of the key themes of
this book – that experimental science is much more than theory testing –
emerges forcefully again; but more importantly, this chapter provides
the first concrete example of the procedures scientists use to bridge the
gap between experimental and nonexperimental circumstances. In the
chapters that follow, then, I try to draw some lessons from this case and
generalize them to build a methodology of external validity inferences.

The market for telecommunication systems

A "decentralizing wave" hit the American economy in the eighties and
nineties. A new political imperative prescribed the replacement, whenever feasible, of centralized, bureaucratic systems of allocation with market processes. Before this wave reached the telecommunications industry,
licenses for wireless Personal Communication Systems (PCS) – providing

161

the right to use a portion of the spectrum for radio communication, telephones, portable faxing machines, and so on – were assigned via an administrative hearing process. Each potential user had to apply to the FCC and convince them of their case; a commission would then decide to whom the license would go. Such a method (the "beauty contest" method, as it is sometimes called) had a number of drawbacks: above all, it was slow, cumbersome, nontransparent, and gave licenses away for free instead of selling them for their market value. In 1982, Congress decided to reform the system and make it faster by using lotteries: each license was randomly assigned to one of the companies that had applied for it.

The lottery system was quicker but had other defects: in particular, companies could participate even if they lacked a genuine interest in the licenses, just to resell them for huge profits later on. A secondary market was thus created in which licenses were resold by lucky winners to those who would really use them. The lottery system generated an unjust and unjustified distribution of income from the controller of the airwaves (the U.S. government) to private individuals who had done nothing to deserve it.[1] In July 1993, Congress decided that the lottery system had to be replaced by a market institution, and the FCC was faced with the problem of identifying and implementing within ten months the best auction system for selling their licenses.

The problem of devising a suitable kind of auction was far from trivial, and previous experiences in New Zealand and Australia had shown that it was also a delicate one: a badly conceived reform could lead to disastrous results.[2] As some journalists put it, "When government auctioneers need worldly advice, where can they turn? To mathematical economists of course," and "as for the firms that want to get their hands on a sliver of the airwaves, their best bet is to go out first and hire themselves a good game theorist."[3] In September 1993, the FCC issued a "Notice of Proposed Rule Making" setting the goals to be achieved by the auctions, tentatively proposing a design, and asking for comments and suggestions from potential bidders.

Soon a number of economists got involved as companies' advisors. The aims to be pursued in auctioning PCS acted as constraints on the

---

[1] According to the U.S. Department of Commerce, more than forty-five billion dollars were gained by opportunistic lottery winners in the eighties (cfr. McMillan 1994, p. 147, n. 3). For a comparison of the auction method of allocation with other methods, such as lotteries and administrative hearings, as well as for a discussion of the pros and cons of auctioning licenses, see McMillan (1995).

[2] McMillan (1994, p. 147) and Milgrom (2004, Ch. 1) tell the stories of these earlier design failures. Later failures are reviewed in Klemperer (2002).

[3] *The Economist*, July 23, 1994, p. 70; quoted in McAfee and McMillan (1996, p. 159).

work of the consultants. The auction was intended to achieve an efficient allocation of the spectrum, to prevent monopolies, and to promote small businesses, rural telephone companies, and minority-owned and women-owned firms (as prescribed by the government and the FCC policy). Moreover, it was understood that the volume of revenue raised by the auctioneer (the FCC) was an important factor to be taken into account. A fairly precise target was thus set right at the beginning of the enterprise. It took the form of an economic phenomenon to be created from scratch, with certain specific characteristics that made it valuable in the FCC's eyes. The following story is a tour de force from this preliminary identification of the target to the final product (the FCC auction as it was eventually implemented in July 1994), through a series of theoretical models, experimental systems, simulations, and public and private demonstrations of the properties of different market designs.

Mechanism design

The FCC enterprise is a typical case of a *mechanism design* problem. The term *mechanism* is widely used in the social sciences, in various and often inconsistent ways. In this chapter, I stick to the meaning that is common in the technical economics literature: mechanisms are *systems of rules* regulating the behavior of agents (individuals, but also institutions such as firms, political parties, etc.) with the aim of achieving certain goals or *outcomes*. The characteristics of the agents (their preferences, beliefs, capabilities, etc.) are usually listed among the properties of the *environment* – the circumstances that are beyond the reach of the legislator and therefore must be taken as given during the process of mechanism design.[4] Mechanism design reverses the form of reasoning that is most

---

[4] There is some ambiguity on whether mechanisms are social entities or representations. Mas-Colell, Whinston, and Green (1995), for instance, on the very same page, define a mechanism as "the formal *representation* of [. . .] an institution" (more precisely: a collection of strategy sets and an outcome function from the Cartesian product of the strategy sets to the set of alternatives), to claim shortly after that "a mechanism can be seen as an *institution* with rules governing the procedure for making collective choice" (p. 866, my italics). The concept of mechanisms is also enjoying some popularity in philosophy of science, in which it is usually defined in close connection with the notion of a system of causal relations or processes; see e.g., Wesley Salmon (1984); Machamer, Darden, and Craver (2000); and Glennan (2002). In an earlier account of the FCC auctions' construction, I tried to unify the economics and the philosophy of science terminology (Guala 2001). Other attempts to apply the philosophy of science idea of mechanism to economics include Pierre Salmon (1998) and his commentators, as well as Dupré (2001). A Ph.D. dissertation by Edward Nik-Kah (unpublished) is the first attempt to write a history of mechanism design theory.

common in science: according to a widely shared view of scientific knowledge, the main task of the theorist is to explain spontaneously occurring and experimental processes by designing an appropriate model for each kind of causal process and the phenomena it generates. The FCC case belongs to an altogether different kind of scientific activity, proceeding in the opposite direction from abstract theoretical models to the concrete systems of rules that govern behavior.

Designs are motivated by a mechanism (a mathematical model, a body of theory) that is perhaps completely devoid of operational detail. The task is to find a system of institutions – the rules for individual expression, information transmittal, and social choice – a "process" that mirrors the behavioral features of the mechanism. The theory suggests the existence of processes that perform in certain (desirable) ways, and the task is to find them. This is a pure form of institutional engineering (Plott 1981, p. 134).

Theory can be used to produce new technology by shaping the social world so as to mirror a model in all its essential aspects. The "idealized" character of the model may thus be a virtue rather than a defect, as the explicit role of theory is to point to a possibility in this case. Theory *projects* rather than describes what is already there.

The mechanisms I am concerned with are market institutions. Mechanism design is often motivated by the will to replace centralized, expensive, or inefficient systems of regulation with "better" (i.e., decentralized, cheaper, or more efficient) ones. An everyday analogy may help here: consider the problem of deciding whether to direct traffic using police officers rather than traffic lights or roundabouts. Each system has advantages and drawbacks in terms of cost, ambiguity of the rules, propensity to generate confusion and mistakes, costs of enforcement, and so on. The theory of mechanism design, then, involves both the study of the functioning of different institutions and their evaluation. It is an enterprise between theoretical and applied economics, which requires first stating clearly the goals to be achieved by the mechanism, and then finding the best means to achieve them given the circumstances. Mechanism design clearly is an activity in which normative and positive economics intersect. Despite the philosophical interest of the normative aspects of mechanism design, in this book, I am concerned with the positive ones only.

Economic theory is obviously an extremely valuable tool in mechanism design. Once the environment (agents' characteristics) is defined, it is possible to think of institutional rules as defining a game, which the agents are facing and trying to solve rationally. Ideally, it should be possible to predict exactly what outcome will be achieved by a given mechanism in a given environment by means of equilibrium analysis. Auction theory, the branch of game theory that is prevalent in mechanism design, is considered "closer to applications than is most frontier mathematical economics" (McAfee and McMillan 1987, p. 700). As it turns out, however, "being close to applications" is not quite the same as "being straightforwardly applicable."

The role of theory

When the FCC started auctioning in 1994, the results were immediately hailed as a major success, if only for the huge sums of money gained by the federal government ($23 billion between 1994 and 1997). Most of the glory went to the theorists who helped design the auction mechanism. The FCC auction was claimed "the most dramatic example of game theory's new power," "a triumph not only for the FCC and the taxpayers, but also for game theory (and game theorists)."[5] It would be a mistake, however, to think of the FCC auction's design as entirely theory driven. Auctions like those for PCS are, in fact, a typical example of what game theory is *not* very good at modeling.

Game theoretic accounts of auction mechanisms date back to the sixties, thanks mainly to the pioneering work of William Vickrey (1961). Vickrey solved an auction game known as the "independent private values model," in which each bidder is supposed to be aware of the exact value of the auctioned item but not supposed to know its value to other bidders. Such an assumption seems to be satisfied in auctions of, for example, antiques, which will be privately enjoyed by buyers who do not intend to resell them. Wilson (1977) and then Milgrom and Weber (1982) extended Vickrey's private value model to other important situations and notably to the "common value" case, in which the exact value of each item is the same for every bidder but unknown to all (auctions for oil leases, for instance, seem to be of this kind). Auctions are modeled as noncooperative games played by expected-utility maximizing bidders. The players are assumed to adopt equilibrium strategies – in the standard sense of a Nash equilibrium in which, given everyone else's moves, no player can do better than she is presently doing by changing her strategy.[6]

---

[5] From *Fortune*, February 6, 1995, p. 36, cited by McAfee and McMillan (1996, p. 159), who also report on other, similar reactions. Other references can be found in Milgrom (1995, 2004).

[6] For an introduction to auction theory, cf. Milgrom (1989); more comprehensive surveys can be found in McAfee and McMillan (1987), Klemperer (2004), and Milgrom (2004).

After more than two decades of theoretical development, the theory of auctions still relies on a number of restrictive assumptions and by no means can be applied to all circumstances. The most important and disturbing feature of the commodities to be auctioned by the FCC (the PCS licenses) is their being, to say it in economists' jargon, sometimes "complementary" (e.g., licenses to provide the same service to different contiguous regions) and sometimes "perfect substitutes" (licenses for different spectrum bands that can provide the same service) for one another. The value of an individual license thus may be strictly dependent on the buyer's owning one or more of the other items: the value of a "package" could differ from the sum of the values of the individual items that are in it. This is the result of a number of characteristics of the airwaves industry, from fixed-cost technology to customer-base development, from problems of interference to the use of different standards by different companies (McMillan 1994, p. 150). For all these reasons, a license for transmitting in a certain region is generally more or less valuable depending on whether one owns the license for a neighboring area or not.

The bulk of auction theory deals with the sale of a single good, and in 1993, the theoretical study of multiunit auctions was still in its infancy. Insights from other parts of economic theory, however, suggested that complementarities could raise great problems, because models of competitive markets with goods of this kind tend not to have a unique equilibrium and tend to be unstable.[7] The theory, at any rate, was highly incomplete: it could not provide a general analysis or a prediction for auctions with interdependent goods. The first issue to be tackled by the consultants when they started work in 1993 was whether to use a traditional bidding mechanism or to create something new, an institution designed "ad hoc" for the specific problem at hand. Although conservative considerations of reliability pulled in the first direction, the peculiar features of the airwaves industry seemed to require the second approach. In the Notice of Proposed Rule Making of September 1993, the FCC suggested the implementation of an auction system in two stages, in which goods are initially auctioned in packages (with a sealed bid mechanism) and, later on, on an individual basis. This procedure seemed to solve the problem of aggregation in a straightforward way: the items are assigned to the winners of either the first or the second auction, depending on which one

---

[7] The literature on multiunit auctions is now growing fast, and recent theoretical developments have mostly confirmed these worries about complementarities (see e.g., Milgrom 2004, Chs. 7 and 8).

is guaranteeing more revenue to the seller. If bidders really value a package more than the individual items in isolation (as the complementarity hypothesis suggests), then the procedure should provide an incentive for aggregation.

"Package bidding" (or a *combinatorial auction design*) was also advocated by the National Telecommunications and Information Administration (NTIA), a public institution advising the government and the FCC on matters of telecommunications policy. Their proposal was based on the work of John Ledyard, a pioneer in the area of combinatorial auctions. Several companies, however, were opposed to selling the licenses in packages, in order to prevent competitors from acquiring a nationwide coverage of the spectrum. Pacific Bell hired two leading auction theorists, Paul Milgrom and Robert Wilson, to advocate an alternative mechanism called the "simultaneous ascending-bid auction"; another economist (Preston McAfee, consulting for AirTouch Communications) also came up with a proposal that differed from the Milgrom-Wilson design only slightly.[8] In a *simultaneous* auction, several markets are open at the same time and bidders can operate in all of them at once. In an *ascending* auction, bidders continue to make their offers until the market is closed – which usually means until no new offers are put forward. Simultaneity and the ascending form allow each bidder to collect valuable information about the behavior of other firms as well as about his or her own chances to construct the aggregation of items he or she most prefers. Bidders can thus switch during the auction to less-preferred combinations as soon as they realize that they will not be able to achieve their primary objective. Moreover, an ascending bid (as opposed to a sealed bid) system is supposed to reduce the risk of "overbidding" (offering more than the value of the item) because by keeping an eye on each other's bids, buyers can make a better conjecture about what the evaluations of the other bidders (and thus the real value of the licenses) are. The proposed mechanism, however, had some worrisome features: in particular, it looked at first sight rather complicated and had never been used before.

On behalf of Pacific Bell, Milgrom and Wilson sent a document to the FCC arguing that a combinatorial process in two stages, like the one

---

[8] The FCC case provides an interesting example of the delicate interaction between scientific and economic interests (the consultants were at times unashamedly lobbying for their companies) in the solution of a techno-scientific problem. Nik-Kah (unpublished) and Mirowski and Nik-Kah (2004) provide an account of the FCC case focusing on political and economic factors, and criticize some aspects of my reconstruction in Guala (2001). On the political context of the auctions, see also Kwerel and Rosston (2000).

Table 8.1. *Payoffs in a Simplified Auction Setting*

Bidder	A	B	AB
One	4	–	–
Two	–	4	–
Three	$1 + \varepsilon$	$1 + \varepsilon$	$2 + \varepsilon$

proposed in the September 1993 notice, was to be discarded. Their arguments are representative of the kind of reasoning used by theorists in order to identify the "right" design for the FCC case. It is worth spending some time reviewing at least one of them, because it sheds light on the role played by game theory in the design process.

Milgrom and Wilson argued, among other things, that a combinatorial institution may give rise to free-riding situations of the following kind (cf. Milgrom 1998, 2000). Suppose there are three bidders (One, Two, and Three). Bidder One's willingness to pay for item A is 4, whereas she is not eligible to buy item B or the package consisting of A and B together. Bidder Two's willingness to pay is symmetrical: 4 for B and not eligible for A and AB. Bidder Three, in contrast, is willing to pay $1 + \varepsilon$ for A, $1 + \varepsilon$ for B, and $2 + \varepsilon$ for AB (with ϵ small and positive). The payoffs are represented in Table 8.1.

The only efficient allocation in this case is the one assigning A to One and B to Two. In an ascending simultaneous auction, Bidder Three bids 1 for each item and Bidders One and Two bid just enough for Three to give up, then they acquire A and B, respectively: the efficient outcome is a subgame perfect equilibrium of the simultaneous ascending mechanism.

Milgrom and Wilson then turn to the two-stage combinatorial design. Under this institutional arrangement, Bidder Three does not have an incentive to bid on A and B individually; she just participates in the auction for package AB, in which she bids $2 + \varepsilon$. Bidder One can win the individual auction for A by bidding 1, and similarly Two can win the auction for B. However, then the package would go to Bidder Three. One and Two therefore have an interest to raise the total value of A and B by bidding more on at least one of them, but would like the other to do so in order to minimize his or her own costs. Milgrom and Wilson show that such a free-rider situation has a mixed-strategy equilibrium that is inefficient. Bidders One and Two each face a subgame that can be represented by means of the payoff matrix in Table 8.2.

By backward induction, it can be proven that this subgame has an equilibrium in which each bidder plays "raise the bid" with probability 2/3

Table 8.2. *Free Riding in a Simplified*
Combinatorial Auction

	Raise Bid	Don't Raise Bid
Raise bid	2,2	2,3
Don't raise bid	3,2	0,0

and "don't raise" with probability 1/3. However, there is a 1/9 probability of Three getting both *A* and *B* by paying just 1/4 of what Bidders One and Two would jointly pay for them (Milgrom 1998, p. 15).

Such an argument clearly makes use of game theory but does not follow from a *general* game theoretic model of combinatorial auctions. Lacking a comprehensive theory of these processes, the economists engaged in the FCC enterprise relied on a number of piecemeal theoretical insights and local analyses of how players are supposed to behave when solving certain tasks in isolation. "The spectrum sale is more complicated than anything in auction theory," as two of the protagonists admitted (McAfee and McMillan 1996, p. 171). The relation between theoretical reasoning and the final implementation of the auction is well summarized by the following remarks:

> The FCC auctions provide a case study in the use of economic theory in public policy. They have been billed as the biggest-ever practical application of game theory. Is this valid? A purist view says it is not. There is no theorem that proves the simultaneous ascending auction to be optimal. The setting for the FCC auctions is far more complicated than any model yet, or ever likely to be, written down. Theory does not validate the auction form the FCC chose to implement. The purist view, however, imposes too high a standard. The auction form was designed by theorists. The distinction between common-value and independent-value auction settings helped clarify thinking. The intuition developed by modelling best responses in innumerable simple games was crucial in helping the auction designers anticipate how bidders might try to outfox the mechanism. (McMillan, Rotschild, and Wilson, 1997, p. 429)

Theory played a crucial role in the design, but only if we interpret the term *theory* in a loose sense.[9] "The auction designers based their thinking on a range of models, each of which captures a part of the issue" (McAfee and McMillan 1996, p. 171). It is true that "game theory played

---

[9] McAfee and McMillan, for instance, claim that "there is a direct link between game theory's laureates and the spectrum auction. The ideas with which Nash, Harsanyi, and Selten are associated – Nash, Bayesian and Perfect Equilibrium – are the basic tools of the theory used in designing the auction" (1996, p. 171, n. 15).

a central role in the analysis of rules" and "ideas of Nash equilibrium, rationalisability, backward induction, and incomplete information [. . .] were the real basis of daily decisions about the details of the auction design" (Milgrom 1995, pp. 19–20), but

[. . .] the real value of the theory is in developing intuition. The role of theory, in any policy application, is to show how people behave in various circumstances, and to identify the tradeoffs involved in altering those circumstances. What the theorists found to be the most useful in designing the auction and advising the bidders was not complicated models that try to capture a lot of reality [. . .]. Instead, a focused model that isolates a particular effect and assumes few or no special functional forms is more helpful in building understanding. (McAfee and McMillan 1996, p. 172)

The other side of the coin is that by reasoning from game theory, it is impossible to define the exact form of the rules to be used in the auction, and theory never gives you the whole picture of the complicated process at any time. For this reason, it is true that "the auctions would not have taken the shape they did were it not for the economic knowledge brought to the design process" (McMillan et al. 1997, p. 429) – but only if we extend the meaning of *economic knowledge* well beyond the theoretical realm. Indeed, today "much of what we know about multi-unit auctions with interdependencies comes from experiments" (McMillan 1994, p. 151, n. 6).

Testbed experiments

A group of experimental economists from Caltech were involved in the design project right from the start. John Ledyard, as already mentioned, was behind the combinatorial design proposed by the NTIA and had a long track record in experiments with multiunit auctions. Another prominent Caltech experimenter, Charles Plott, was hired by Pacific Bell in 1993 to run a series of experiments that would test some key theoretical insights and to help choose the best auction design. Later, Plott would also play a prominent role in the careful checking of the rules and software to be used in the first real FCC auction.

Initially, the role of experiments was to discriminate among different institutions (the "hybrid" design proposed by the FCC, the Ledyard combinatorial auction, and the Milgrom-Wilson-McAfee "continuous ascending" auction), for which neither a comprehensive theoretical analysis nor empirical data existed. The economists involved in the design process quickly focused on efficiency as the main objective to be achieved by the auctions, and *efficiency of the spectrum* was interpreted (somehow twist-ing the original FCC ruling) in strictly economic terms as "assigning the licenses to the bidders who value them most."[10] By theoretical means, it is impossible to prove that the allocation generated by the continuous ascending auction is efficient. By observing a "real world" auction, very few hints about whether the outcome is efficient or not can be gained because of the unobservable nature of bidders' valuations. Experiments, in contrast, allow the "induction" of known values on subjects and thus enable one to check whether a given institution really allocates the goods to those who value them most. The Caltech team made use of what they call "testbed" experiments:

An experimental "testbed" is a simple working prototype of a process that is going to be employed in a complex environment. The creation of the prototype and the study of its operation provides a joining of theory, observation, and the practical aspects of implementation, in order to create something that works. (Plott 1996, p. 1)

In early 1994, the Caltech group ran a series of comparative efficiency tests of the simultaneous ascending auction versus a combinatorial sealed bid and the hybrid FCC design. The latter emerged as the least-performing format, whereas the combinatorial design came out on top in experiments with strong complementarities. Ledyard, however, recognized that the implementation of a combinatorial auction raised too many complicated issues to be tackled in the tight time frame provided by the government (Kwerel 2004, p. xx), and therefore the simultaneous ascending design emerged as the winner due to a combination of empirical and practical reasons.

In this first stage, experimenters were particularly interested in observing the functioning of the two bidding institutions in their "bare-bones" versions, in order to become familiar with their fundamental prop-erties and problems. Comparative testing of institutions in their simplest versions, however, was also instrumental in letting the first operational details and problems emerge. The mechanisms were for the first time transported from the abstract world of ideas to the laboratory. Some "flesh" was added to transform the fundamental insights into working

[10] The history of the other objectives initially pursued by the FCC has been thorny: a 1995 Supreme Court ruling induced the FCC to eliminate positive discrimination on the basis of race and gender (McAfee and McMillan 1996, p. 167). Other forms of discrimination (in favor of small businesses and rural telephone companies) remain, but their success remains controversial.

processes, flesh that might have modified, impeded, or hidden some of the systems' structural dispositions. These experiments could therefore be seen as tests of the various designs – but "tests" of a peculiar kind, because specific versions of the auction systems themselves, rather than some theoretical model of the auctions, were subject to examination. The issue was not whether some theory about an institution was to be discarded, but whether one rather than another institution should have been chosen for its properties, despite the fact that no thorough theoretical understanding of its functioning was available. Given the effect to be achieved, in other words, a mechanism had to be discovered or constructed that could deliver it reliably.

At the first step of "concretization," unsurprisingly, experimentalists spotted problems that the theorists could not have anticipated. Plott and his collaborators implemented the FCC proposal by running so-called Japanese auctions combined with a sealed bid preauction. In the first round, a sealed bid offer is made for a package of items, which are then auctioned individually in the second round: in a Japanese auction, the auctioneer announces higher and higher prices and the bidders "drop out" one after another; the last one to remain in the auction wins the item and pays the second highest bid. If the aggregate value of the individually sold items exceeds that of the package, the results of the sealed bid auction are disregarded; otherwise, the items are sold as a package.

One problem with this procedure is that some bidders have an interest in staying in the Japanese auction well above their reservation price, in order to raise the prices and overcome a sealed bid preoffer. This may push the price "too high" and cause a bubble effect. The risk of staying in the auction above your reservation price, however, increases as the number of bidders who are participating diminishes. For this reason, information is crucial: players must not know how many others are still competing. Not to communicate explicitly how many bidders are still in the game, however, does not constitute a sufficient form of shielding: one has to be careful that *any* possible information flow be stopped. The click of a key, a "blink" on the computer screen coinciding with bidders' dropping out, or even a door slamming when they leave the room is enough to give valuable hints to the other participants. These problems were actually discovered by Plott and his team thanks only to laboratory work, and the Japanese auction design was abandoned also because of these practical difficulties.

The problem above is one of "robustness." The FCC proposal, in other words, is not only outperformed by the continuous ascending auction in

terms of efficiency, but is also difficult to implement correctly. It is a delicate process, and small imperfections (e.g., in shielding the flow of information) may cause it to break down altogether. One can distinguish cases
of "environmental" robustness like the one above from cases of "personality" robustness (Schotter 1998): although economists' assumptions in
modeling economic agents for the most part may be accurate, in more
or less marginal instances, real behavior may diverge from the model's
predictions. Environmental robustness, in other words, is a function of an
institution's capacity to work properly in a number of different environments. Personality robustness, in contrast, depends on its capacity to work
with a range of real agents, who may behave slightly differently from the
ideal rational maximizing agents postulated by game theoretic models.

Confusion and misunderstanding of the rules are also sources of concern in the implementation of mechanisms, not the least because they
may result in litigation. Testbed experiments allowed the identification of
those critical moments when subjects may need help from the auctioneer in order to understand some detail of the auction. The problem is
not just whether a certain abstract model fits reality (or whether reality
can be made to fit the model), but also *how* it fits. The question is how
fragile such a fit is and how sensitive it is to little imperfections, mistakes,
and so on.

Another important role of laboratory experiments consisted of helping to develop the appropriate software. Game theory, of course, does
not specify whether an auction should be run with mere pencil and paper
support or with computers. Electronic auctions are generally privileged
because they facilitate the analysis of data and the enforcement of the
rules: one can, for instance, design software that does not accept bidding
below the highest standing bid of the previous round, thus automatically
controlling for errors and saving precious time. However, no specific software for the continuous ascending auction was available at the time, and
a new program had to be created ad hoc. The data of testbed experiments,
elaborated by means of an independently designed program, were used as
inputs in the final software to be used in the real FCC auctions. By means
of such "parallel checking" (Plott 1997, pp. 627–31), the consultants made
sure that the FCC software worked properly and delivered correct data
in a reliable fashion. Trained students were employed to investigate the
properties of the software. They used diaries and notebooks to keep track
of all the problems that could arise and then were asked to answer a questionnaire surveying the most likely sources of "bugs." The questionnaire
included questions like,

What happens if you stay logged on after the initial withdrawal; What happens if you log in from multiple locations at the same time; what happens if you enter 0000 rather than 0; what happens if you are theoretically inactive but then you log on after various events; what happens if you log in at the last second of a session or have a power failure; what happens if you follow local software installation exactly to the "letter" of the instructions; etc. (Plott 1997, p. 631, n. 4)

The process must be "idiot-proof": robust to the people who use it, who may create a number of bizarre problems (up to crashing the whole network) or make unpredictable moves in playing their game.

The enterprise of building a "properly working" institution in the laboratory thus shows that small variations can make a big difference indeed. "The exact behaviour of the auctions can be sensitive to very subtle details of how the auction process operates" (Plott 1997, p. 620; see also Klemperer 2004). To understand why a mechanism may be chosen for reasons that differ from its theoretical properties, one has to remember that those properties are defined at a certain (usually high) level of abstraction. An institution producing desirable allocations when correctly built may nevertheless be very difficult to build correctly. In the context of the natural sciences, Nancy Cartwright calls it the *problem of material abstraction*:

A physicist may preach the principles by which a laser should operate; but only the engineers know how to extend, correct, modify, or sidestep those principles to suit the different materials they may weld together to produce an operating laser. (Cartwright 1989, p. 211)

Compared with other applied disciplines, economic engineering is peculiar in at least two respects: first, once the basic causal structure has been chosen (e.g., the combinatorial auction design), the economic engineer has very little room for maneuvering in terms of the corrections, modifications, and interventions that are allowed. The materials are, for the most part, given because agents should not, for instance, be forced to change their preferences – although some kinds of preferences (e.g., collusive or altruistic ones) can be neutralized by design. The principal way in which the economist can intervene on the problem-situation is by defining the rules of the institution. Second, as we have seen in the FCC case, the designers were short of theory guiding the implementation. It was not just a matter of "adding matter" to an abstract causal structure, but rather of adding matter while understanding and shaping the causal structure itself.

Testing the rules

The Caltech experiments were used differently in different phases of the project (Plott 1997, p. 608). First, as we have seen, they were instrumental in choosing between the Milgrom-Wilson-McAfee design and the initial FCC proposal. Second, testbeds were used to transform the abstract design into a real process that could perform the required task reliably in the environment in which the auction was to be implemented. Finally, experimental data were most valuable in interpreting and checking the results of the first *real* auctions run by the FCC. Before coming to that, however, it is worth looking at some of the experiments run by Plott and his team in order to check the joint effect of the rules that would have regulated the FCC auctions.

Mechanism designers see the rules as a device for defining the strategic situation that bidders face and to which they are supposed to react rationally. Unfortunately, in the case of spectrum licenses, theorists were unable to solve the game as a whole and therefore had to rely on the analysis of single subgames in a piecemeal manner. How the pieces would have interacted once put together remained an open issue, about which theorists had relatively little to say. Experiments can be used to move gradually from the world of paper and ideas to the real world, without losing the desired structural properties of a mechanism along the way. The enterprise is similar to that of designing a space probe: it would be too costly to proceed on a trial-and-error basis and perform a series of full-scale tests.

The structure of the continuous ascending auction so far has been sketched at a most abstract level. The details of the design are, in fact, much more complicated. Preston McAfee, Robert Wilson, and Paul Milgrom were mainly responsible for writing the detailed rules that were to regulate bidding in all its various aspects and that eventually were put together in an official document. The most important – and debated – rules concerned increments, withdrawals, eligibility, waivers, and activity.

The simultaneous ascending auction proceeds in rounds. At every round, the players offer their bids – which are scrutinized by the auctioneer – and then are presented with the results. The feedback includes the bids presented at that round, the value of the "standing high bid," and the minimum bid allowed for the next round. The minimum allowed bid is calculated by adding to the standing high bid a fixed proportion (five percent or ten percent, usually) called the *bid increment*. A bid is said to be eligible if it is higher than the minimum bid allowed in that round and it is presented by an *eligible bidder*. Each bidder, in fact, must at the beginning of the auction make a deposit proportional to the number of licenses she wants to compete for (each item is auctioned on a different market). Such a deposit establishes her "initial eligibility," that is, the number of markets she can enter.

The idea of eligibility was introduced not only to prevent problems such as those that occurred in the New Zealand auctions (in which bidders who were not really interested in the licenses participated and then resold the licenses to others), but also to regulate the duration of the auction. Eligibility constrains a bidder's activity in an ingenious manner. A bidder is said to be *active* if she either has the standing high bid from the previous round or is submitting an eligible bid. The activity cannot exceed a bidder's initial eligibility, but bidders also have an incentive not to remain below a certain level of activity. In the first round, in fact, a buyer must bid at least a certain fraction of her initial eligibility; if she does not, her eligibility is reduced in the next round. Such a procedure is supposed to increase the transparency of the auction (by forcing companies to commit early on), and to speed it up by preventing "wait and see" strategies. The possibility of mistakes is taken into account by providing bidders with five *waivers* of the activity rules. Bidders could also *withdraw* a bid, but with the prospect of paying the difference between the final selling price as elicited in a further round and the withdrawn bid, in case the latter exceeded the former.

The rules regulating activity were motivated by the worry that bidders could have a strategic interest in slowing down the pace of the auction, and that the auction could even never end at all.[11] The duration of an auction depends on two factors: the number of rounds played and the interval between rounds. According to the Milgrom-Wilson-McAfee design, each auction is supposed to stop after a certain period and start again the next day, until no new eligible bids are received. The idea of having subsequent "rounds" was motivated by the thought that companies may need time to revise their strategies, reflect on their budgets, and thus avoid expensive mistakes. As Paul Milgrom put it, "There are billions of dollars at

---

[11] Milgrom (1998) provides an example and a game theoretic argument in support of such a hypothesis. The worry about never-ending bidding was motivated by the choice of the Milgrom end rule, i.e., that an auction should be considered closed only when no bidding is taking place in *any* of the simultaneous markets. McAfee had originally proposed to close bidding independently in each market, but this would have made the creation of valuable packages more difficult.

stake here, and there is no reason to rush it when we are talking about permanently affecting the structure of a new industry."[12]

The FCC, however, was concerned about time, not least for practical reasons (booking a big hotel to run an auction, for instance, costs money). Whereas theory does not tell you how long a continuous ascending auction may go on, experiments allowed the testing of different rules, with rounds taking place at different intervals. One possible solution was to impose big increments above the highest standing bid so as to identify the winners quickly. In experiments, however, it was observed that big increments sometimes eliminated bidders too quickly, causing their eligibility to drop and therefore creating a "demand killing" effect (Plott 1997, p. 633). Such an interaction between the increment rule and the eligibility rule could have hardly been predicted without experiments. Without a general theory of simultaneous ascending auctions, theorists could rely only on a number of independent insights about the effects of different rules but could not exactly figure out what would happen were all the rules implemented at the same time.

The concepts of withdrawals, eligibility, increments, and announcement of stage changes, all involve reasonable sounding concepts when considered alone, but there remain questions about how they might interact together, with other policies, or with the realities of software performance. Can one waiver and bid at the same time? What happens if you withdraw at the end of the auction – can the auction remain open so the withdrawal can be cleared? How shall a withdrawal be priced? (Plott 1997, p. 629)

The answers to these questions were sought partly in the laboratory. "The complex ways the rules interact, and the presence of ambiguities, do not become evident until one tries to actually implement the rules in an operational environment" (Plott 1997, p. 628). More time between rounds might allow bidders to put forward sensible bids, but more frequent rounds might also shorten the process considerably. And would either of these solutions affect efficiency? Plott and his collaborators eventually found that total time was principally a function of the number of rounds, and auctions going on for more than one hundred rounds were observed in the laboratory. The Caltech team thus tried to vary the interval between rounds and concluded that having more frequent rounds did not affect efficiency in their laboratory experiments (Plott 1997, pp. 632–3).

---

[12] "Access to Airwaves: Going, Going, Gone," *Stanford Business School Magazine*, June 1994.

The Milgrom-Wilson-McAfee rules also involved the possibility of withdrawals, in case a winner decided a posteriori that the price was not worth the item bought or that she could not achieve the preferred aggregation: the item would be "sold back to the market" and the withdrawing bidder would pay the difference between her bid and the final price to the FCC. Withdrawals and losses may, however, cause "cycles": an item may be bought and resold, bought and resold, and so on until someone is satisfied with the price. Experiments were used to create cycles in the laboratory (see Figure 8.1) in order to see whether they could arise in practice, to study the conditions (the parameters) under which they are generated, and how they behave (e.g., for how long they go on before disappearing).

To sum up, theoretical and empirical insights initially provided just a few rough "stories" about the functioning of certain bits of the institution. Thanks to previous experiments, it was known that more transparent mechanisms (like the English system) tend to prevent winner's curse effects better than less transparent ones (e.g., sealed bid auctions). It was also known that imposing no entry fee would have encouraged opportunistic bidders to participate. It was conjectured that under some circumstances, some bidders might have an interest (opposite to the FCC's) to delay the auction – and so on. These insights do not jointly pin down the structure of the best possible institution in all its aspects. They just convey information about what certain components do in certain cases, when acting in isolation. Instead of laying down the structure of the auction on paper, it was displayed at work in the laboratory. The job of checking that the different components had been put together in the "right" way (that they *could* be put together in the right way, to begin with) and that they worked smoothly was done by running demonstrations in the lab.

Checking external validity

When you build a new technology, you do not just pretest it accurately – you also *check* that it's doing its job properly after it has been implemented. The FCC consultants, in fact, monitored a real auction (which took place in Washington, DC, in October 1994)[13] to check whether the

---

[13] For a detailed analysis and discussion of the early FCC auctions, cf. Cramton (1995, 1997, 1998), Ayres and Cramton (1996), and Milgrom (2004). The data from all the auctions run by the FCC are available at http://wireless.fcc.gov/auctions/.

transposition of the simultaneous ascending auction mechanism from the artificial world of the laboratory to the real world had been successful. The expertise achieved in the lab was extremely valuable for this purpose. Experiments had taught that, especially in the first rounds, "regardless of the amount of preparation and testing, things happen," and "decisions must be made on the spot from experience and judgement" (Plott 1997, p. 631). A committee of consultants was thus formed to supervise the FCC auction and intervene at any time to assist the participants. Any inconvenience that arose in testbeds was likely to be spotted by the trained eye of the experimenters.

Laboratory pilots with parameters similar to those expected in the real auction were run beforehand so that the results could be compared after the event. The experimental auctions had been constructed so as to have the same number of players, the same number of items auctioned, complementarities such as those presumably existing in the "real" market, a similar number of rounds,[14] similar (although fictional) bids, and so on. Then, a large amount of data collected both in the lab and in the real auction were systematically analyzed and compared. This data included bidding patterns, pricing trajectories, the rise of "bubbles," the formation of sensible license aggregations, the fact that similar items sold for similar prices, evidence of postauction resale, and several other phenomena of interest (see also Cramton 1995, 1997, 1998). To convey the taste of this procedure, the "equilibrating" trajectory of an experimental testbed auction is represented in Figure 8.1.

The dots on the curve stand for the revenues obtained in the laboratory auction, whereas the horizontal line represents the revenues predicted by the competitive equilibrium model, computed on the basis of experimentally induced demand and supply curves. The adjustment process toward equilibrium is one of the most replicated effects in experimental markets: it takes time and repetition to "find" equilibrium prices. In Figure 8.2, the revenue of the real FCC auction of October 1994 is represented.

The path is remarkably similar to that of the experimental auction. (The downward sloping curve represents so-called excess bids and can be ignored for our purposes.) The price trajectories in the two auctions evolved in a similar way, and because experimenters *knew* that the laboratory auctions delivered approximately efficient outcomes in the long run,

---

[14] The FCC, though, interrupted the final testbed experiments before their completion. One last round of bidding was unexpectedly called, and this probably accounts for the sudden rise in prices observable in Figure 8.1 (see Plott 1997, p. 636, n. 6).

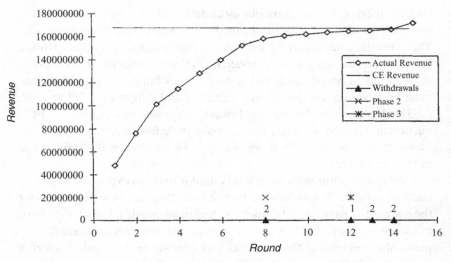

Figure 8.1. Price trajectory in a laboratory auction. Reproduced by permission from Plott 1997, p. 634. © Blackwell Publishing.

they felt confident in arguing that an efficient allocation was achieved in the Washington auction as well. The reasoning behind such a conclusion is straightforward:

> If indeed the same principles were operating in the FCC environment then the FCC auction converged to near the competitive equilibrium and exhibited high efficiency. (Plott 1997, p. 637)

The external validity argument takes the form of an inference from observed data and a set of background assumptions to the underlying generating process that produced the data. The structure of such an inference can be reconstructed as follows:

1. If all the directly observable features of the target and the experimental system are similar in structure;
2. If all the indirectly observable features have been adequately controlled in the laboratory;
3. If there is no reason to believe that they differ in the target system;
4. And if the outcome of the two systems at work (the data) is similar;
5. Then, the experimental and target systems are likely to be structurally similar mechanisms (or data-generating processes).

The strength of this argument lies in the work the FCC and its consultants had done to ensure that the processes that took place in reality

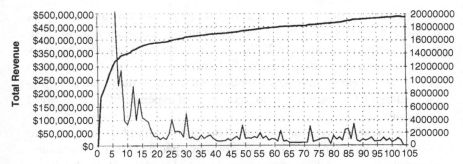

Figure 8.2. Bidding trajectory in the FCC auctions. Reproduced by permission from Plott 1997, p. 636. © Blackwell Publishing.

were the same as those they had observed in their laboratory. The same causes are supposed to operate because experimenters built the two systems so as to be structurally similar to each other. The transportation of the mechanism outside the laboratory was as smooth, gradual, and carefully monitored as possible. The bootstrapping inference was based on a ground of theoretical, practical, and experimental knowledge.[15]

Some lessons from economic engineering

The FCC auctions were in many obvious ways a success for the FCC and its advisors. They were successful politically, to begin with,[16] by raising unexpected revenues for the government. Game theorists established themselves as scientific consultants, and the simultaneous ascending design has since inspired other projects as a benchmark example of economic engineering.[17] This does not mean that the implemented institution was

---

[15] Notice that I'm focusing here only on the arguments for efficiency that make use of experimental data. For more evidence and analysis of the first auctions, see McAfee and McMillan (1996), Cramton (1995, 1997, 1998), and Ayres and Cramton (1996). A vocal opponent of the orthodox interpretation of the auctions' "success" is Murray (2002); see also Mirowski and Nik-Kah (2004). Undoubtedly, because of the difficulty of proving beyond doubt the efficiency of the auctions, many apologists end up identifying revenue as the main criterion of appraisal (contrary to the original spirit of the FCC enterprise).

[16] U.S. Vice President Al Gore claimed that "now we are using the auctions to put the licenses in the hands of those who value them the most" (quoted in Milgrom 1995, p. 1).

[17] See, for example, the auctions for the third generation of mobile telephony run in several European countries in the past few years (Klemperer 2002, 2004; on the UK auctions in particular, see Binmore and Klemperer 2002). Nothing has been said in this paper about the business aspects of mechanism design, but there is certainly an interesting story to be told about game theory turning into a profitable enterprise (Nik-Kah unpublished and

flawless: some work has been and is still being done on various aspects of the design. Collusive behavior, for example, has been a major concern for some years, and much effort has also been invested in the study of combinatorial package bidding.[18] However, these developments go beyond the scope of this book. I am interested here in the methods implemented by FCC consultants and in the role of experiments in real-world intervention.

The FCC auctions belong to a small but growing number of attempts to create new market institutions to perform prespecified tasks. Other examples include the matching mechanism designed by Alvin Roth (Roth and Peranson 1999) to regulate the job market of graduates in medicine, and the electric power market designed by Robert Wilson (2002) in California. Experimental knowledge of individual decision making is also applied to help improve the decisions of professionals – doctors, for example – who face difficult (and potentially costly) decisions in their everyday work. The "debiasing" program, as it is usually called, makes extensive use of the results of experiments devoted to testing rational decision theory and of the "robust" biases that have been discovered therein (cf. Fischhoff 1982; Gigerenzer, Todd and the ABC Research Group 1999).

These projects have opened new, unexpected possibilities. Economists have been traditionally concerned with the analysis of existing, "naturally evolved" entities. Working in such conditions made them look more like meteorologists than physicists. The market design revolution in the 1990s teaches an important lesson: it is difficult to understand a system (physical or social) unless you can intervene and experiment with it. And it is even more difficult to predict its future behavior, unless the system has been shaped and disciplined so as to work "appropriately." Natural scientists learned this lesson a long time ago. Philosophers of science have been slower to appreciate its implications, possibly because they have been working within a framework that does not draw the appropriate distinctions between laboratory and nonlaboratory science – but exactly what methodological implications should be drawn from the study of applied science?

Mirowski and Nik-Kah 2004 strongly advocate this reading of the FCC story). Several economists involved in the FCC project later created consultancy firms, and it has become customary nowadays to patent new auction systems.

[18] Some companies, for example, devised sophisticated signaling systems using the last few digits of their bids in order to communicate to other competitors their willingness to collude (Weber 1997, Cramton and Schwartz 2000). On combinatorial auctions, see Ledyard, Porter, and Rangel (1997); Bykowsky, Cull, and Ledyard (2000); Milgrom (2004, Ch. 8), and the FCC website at http://wireless.fcc.gov/auctions/conferences/.

So far we have examined one particular example, and it would be premature to derive strong implications from this case only. Some features of the FCC case, however, are worth emphasizing: first, the various roles played by experiments at different stages during the project. Whereas experiments in the early stages investigated the functioning of mechanisms at a rather abstract level of specification, a strong inference about the behavior of the real FCC auctions was based on the evidence collected in experiments that replicated real-world conditions rather closely (albeit necessarily always imperfectly). Similarity between the laboratory and the real world is therefore not necessary for experiments *in general*, but is nevertheless important when one is aiming at making strong external validity inferences about a specific environment.

Second, we have seen how the external validity step was based on a comparison between laboratory and real-world evidence, and stood on a stock of "background knowledge" aimed at making the inference as strong as possible. This is in line with what has been said in the first part of the book, but of course is still a rough account of inferences from the experiment and has to be substantiated by means of more concrete examples and methodological analysis. What sort of conditions must be in place in order for the inference to be strong? Is such an inferential procedure restricted to the lucky case in which we can shape the real world so as to mirror laboratory conditions, or can we generalize to other, less fortunate circumstances? The next two chapters are devoted to answering these questions.

NINE

From the Laboratory to the Outside World

Mechanism design is applied science carried out under the most favorable conditions – in conditions that maximize the chance of successfully exporting experimental results outside laboratory walls. In this chapter, I try to generalize from this particular case to provide an account of external validity inferences in less than ideal circumstances. Scientists, as a matter of fact, do draw inferences from experiments, even when the outside world cannot be shaped to resemble laboratory conditions. I introduce a new example from auction theory in which the procedure is exactly symmetrical to the one followed by the FCC consultants: the laboratory conditions were shaped so as to resemble those of a real-world economy, instead of the other way around. I also draw some analogies from experimental medicine and elaborate on the role of laboratory and field evidence in external validity inferences. Before doing that, however, I have to dispose of a disturbing and extreme stance on the external validity problem, which I label "radical localism."

Radical localism

The cathodic rays that make our TV sets function are generated by a carefully engineered causal structure that has been repeatedly tested in laboratory conditions, then in the factory, and finally stabilized, "shielded" within a plastic box, and sold to customers all over the world. With few exceptions (e.g., Cartwright 1999), philosophers of science have neglected this pervasive feature of applied science (or "techno-science," as it is sometimes called): success is usually achieved as much by construction as by accurate representation. We can control and predict what happens under circumstances X not only when our models (theories, hypotheses) fit X, but also by shaping X to fit our models. In the late 1970s and 1980s, a new breed of historians and sociologists of science began to highlight the connection between scientific knowledge and technology building in their studies of experimental practice.[1] Bruno Latour and Steve Woolgar's *Laboratory Life* (1979) is one of the pioneering works in this literature. It focuses on the research on thyrotropin-releasing hormone (TRH) that led to the award of the Nobel Prize for medicine to Roger Guillemin in 1977.

TRH is a product of the mammalian hypothalamus that contributes to the production of a hormone called thyrotropin. The hypothalamus secretes only a tiny amount of TRH, which is therefore very hard to obtain. Guillemin (and, independently, Andrew Schally) bypassed the scarcity problem by synthesising a tripeptide whose behavior very closely resembles that of TRH. Latour and Woolgar describe in detail the process of identification of the synthesized protein with the real TRH in mammals' hypothalamus. The remarkable aspect of this case, as noticed by Hacking (1988), is that the artificial peptide became quickly – and without much scrutiny – the benchmark for deciding what should count as TRH and what should not. Its structure became *the* structure of TRH. The success of Guillemin's and Schally's work was probably the result of their ability to provide an off-the-shelf substance of vast potential applicability (independently of its being *the* original TRH or not). However, in several passages of their book, Latour and Woolgar go further than that: they suggest that one cannot rationally establish that an artificial result in the laboratory is actually a faithful representative of a real-world phenomenon or entity. Scientific knowledge never really trespasses laboratory walls, despite scientists' rhetorical claims.

This thesis is endorsed more explicitly in Latour's later work. In *The Pasteurisation of France* (1984), Latour reconstructs Luis Pasteur's struggle to enroll the scientific community in his research program of experimental microbiology and extensive vaccination. Latour tries to explain Pasteur's success, a success that – he believes – went well beyond what was justified by common or reasonable scientific standards. Latour takes sides with the dissenters, such as Koch and Peter, who criticized Pasteur's "hasty generalization" from a bunch of vaccinated sheep to "a *general* method, applicable to all infectious diseases" (Latour 1984, p. 29). The skeptics

---

[1] Cf. e.g., Latour and Woolgar (1979), Collins (1985), Latour (1987), Galison (1987), Gooding (1990), and Pickering (1995).

questioned the stock of empirical facts upon which Pasteur based his inferences. According to Latour, "no one can deny that in 1881 this stock was extremely limited" (1984, p. 30). What Pasteur had done, in other words, did not support the generalized theories of infection and cure that he in fact formulated. The upshot is that Pasteur's "generalizations" were just rhetoric. In reality, Pasteur simply "exported" his findings by shaping reality on the model of his laboratory on the Rue d'Ulm. Where this was not done properly, the "generalizations" failed.

It often seemed for instance, that the antianthrax vaccine refused to pass the Franco-Italian border. However much it tried to be "universal," it remained local. Pasteur had to insist that the practices of his laboratory be repeated exactly if the vaccine were to travel. (Latour 1984, p. 93)

Pasteur used to perform dramatic public displays of his scientific achievements. In May 1881, for example, he demonstrated the effectiveness of his vaccine on a small farm in the village of Pouilly-le-Fort.[2] Latour takes the experiment at Pouilly-le-Fort as a paradigmatic case of external validity inference: its successful outcome was achieved by turning the farm into a laboratory. Where reality has not been carefully engineered, according to Latour, laboratory results do not apply. Experimental results travel from lab to lab, but never really come to grips with the outside world. Here are a couple of quotes, in typical Latourian aphoristic style:

When people say that knowledge is "universally true," we must understand that it is like railroads, which are found everywhere in the world but only to a limited extent. To shift to claiming that locomotives can move beyond their narrow and expensive rails is another matter. Yet magicians try to dazzle us with "universal laws" which they claim to be valid even in the gaps between the networks. (1984, p. 226)

Whatever is local always stays that way. (1984, p. 219)

I will call *radical localism* the view that experimental results do not apply to the world outside the laboratory.[3] Radical localism is disturbing: it would be disappointing to find out that biomedical researchers experiment with drugs on animals but will never be able to tell whether these drugs can cure us (humans).[4] And similarly, we would be disappointed

---

[2] The original report of the experiments can be found in Pasteur (1881).

[3] Cf. also Latour (1987, pp. 247–54). David Gooding (1990, Ch. 6) and Andy Pickering (1995) at times seem to defend similar views.

[4] Not to mention the fact that animal experimentation surely would be morally outrageous if it were not able to help in the cure of human beings. See LaFollette and Shanks (1995) for a critique of animal experimentation along these lines.

to find out that economists' experiments can teach us nothing about the working of real-world economic systems. Radical localism must face a normative challenge: as worldly decision makers, we *require* experimental knowledge to extend outside the laboratory.

In order to challenge its disturbing aspects, one can question the descriptive adequacy of radical localism. *Some* science, to be sure, travels from lab to lab without ever being faced with unconstrained reality. But not *all* science works that way, and indeed scientific knowledge would be a poor thing if it were limited to that. Many cases of scientific application undoubtedly fit the radical localist account: the FCC auctions, as we have seen, were designed and accurately tested in the economic laboratory at CalTech before being exported to the real world. They did not exist as a "naturally evolved" entity before the experiments took place. However, exporting the laboratory is just one route toward external validity – the safest one perhaps, but not the *only* one. Its viability depends on how much we are allowed to intervene and shape reality to fit our experimental prototypes. In many cases, we are not allowed to do that for ethical, political, or merely practical reasons. In such cases, scientists have to follow some alternative strategy – but what kind of strategy?

Laboratory and field evidence

The core intuition behind localism is that controlled experimentation and cautious inference from the evidence are the key recipes for success in applied science. Contrary to the standard view, broad generalizations and universal theories are often the enemies of scientific progress. It is easy to find examples of that. In 1917, the discoverer of vitamins A and B, Elmer McCollum, reported that experimental rats did not seem to suffer from scurvy when put on the same diet that produces scurvy in guinea pigs (McCollum and Pitz 1917). McCollum subscribed to the view that the nutritional process of all single-stomached mammals must be the same. He was testing the law, "For all single-stomached mammals, if X then Y" (where X is a dietary regime, and Y is scurvy). But in rats, X and not Y, which logically falsifies the general law (by *modus tollens*). McCollum rejected the hypothesis that scurvy results from a nutritional deficiency, an idea defended only a few years earlier by Axel Holst and Theodor Frölich (1907) on the basis of their experiments on guinea pigs.

In order to explain Horst and Frölich's results away, McCollum examined several dead guinea pigs and noticed an accumulation of "putrefying" feces in a tract of the intestine. He therefore concluded that scurvy must be caused by constipation and the transmission of bacteria from the intestine. Some attempts to cure animals with laxatives gave mixed results until about a year later, when research teams in London and New York noticed that all guinea pigs (regardless of whether they died of scurvy or not) accumulated putrefying feces in the intestine. They also, surprisingly, noticed a strong *positive* correlation between consumption of raw milk and constipation, and a strong *negative* correlation between consumption of milk and scurvy (Carpenter 1986, pp. 182–3). Animals fed big quantities of raw milk, in other words, were most likely to be constipated, but were also less likely to develop scurvy. This ruled out McCollum's "intoxication" hypothesis in the laboratory, but the same negative correlation between milk and scurvy was also observed in a sample of young children. This finding paved the way for further research on the properties of milk and other scurvy-preventing food, which eventually led to the discovery of vitamin C.

We find here some typical features of successful experimental science. First, a result is proven in a specific, somehow idiosyncratic experimental system (laboratory animals); then it is generalized using data from the experiment and from the field (human patients, in this case). Latour agrees that successful science requires such a combination of laboratory and field evidence. In *The Pasteurisation of France*, he uses the metaphor of a "translation": from real-world phenomena to laboratory phenomena, and then back from the laboratory to the field. The first stage is one of fact gathering: it is necessary to "learn from people on the ground – farmers, distillers, veterinary surgeons, physicians, administrators – both the problems to be solved and the symptoms, the rhythm, the progress, the scope of the diseases to be studied" (Latour 1984, p. 76). Gathering facts is functional not only to a proper understanding of the disease, but also to achieving the final effect of external validity, in the third stage. Pasteur's predecessors, the "Hygienists," had been unable to explain one important feature of infectious disease – its *variation*, the strange, uneven patterns that it could take: why a disease came up in one place but not in another, why now but not then, why a certain group of people was affected but not another, why certain animals but not all animals of that kind (cf. Latour 1984, p. 22 and pp. 63–4). In order to convince the scientific community and government officials, Pasteur had to explain the variation in contagiousness, and he did so by reproducing it in the laboratory: Pasteur "in effect simulated an epidemic" (1984, p. 63). He did it in his laboratory, where the disease could be controlled, manipulated, and measured at will.

This is the second, crucial stage of the translation. Pasteur created an "'experimental illness,' a hybrid that had two parents and was in its very nature made up of the knowledge of the hygienists and the knowledge of the Pasteurians" (1984, p. 63). The laboratory disease then is a "mediating" entity, an independent object whose relevance to the problem at hand cannot be taken for granted but has to be demonstrated empirically. Pasteur himself recognized that a further step has to be taken:

These are still laboratory experiments. We must find out what happens in the countryside itself, with all the changes in humidity and culture. (Pasteur 1922, VI, p. 259; quoted in Latour 1984, p. 76)

In the third stage, then, Pasteur moves back to the field, taking his laboratory knowledge and apparatus with him. Latour here stresses two points: Pasteur must retain all the power of the laboratory (the efficacy of his scientific tools) but also compromise with the environment. The field is transformed so as to achieve the first desideratum, but the laboratory setting must be accommodated too, to make the tools applicable to a larger-scale problem. The moral seems to be the following: if you cannot export some laboratory conditions into the real world, you had better make sure that the relevant aspects of the real world are imported into the lab. It is now time to see the strategy at work in an economic experiment.

Mimicking the real world

In 1971, the Atlantic Richfield Company claimed that the low profits from the exploitation of oil leases in the Outer Continental Shelf (OCS) were the result of a so-called winner's curse phenomenon (Capen, Clapp, and Campbell 1971). Auctions of this kind are "common value auctions" – auctions in which the value of the auctioned item is the same for all participants but initially unknown to all. A crucial part of the bidding game, then, consists of trying to estimate the true value of a lease. When the participants fail in this estimation, the winning bid is likely to turn out to be overoptimistic and the exploitation of the lease unprofitable (the winner is "cursed").[5]

The claims of the Atlantic Richfield Company were suspect: the company clearly had an interest in convincing other competitors to be more cautious in their valuations, and their claim could be read as a disguised

---

[5] For an introductory survey of the literature on the winner's curse phenomenon, see Thaler (1988).

invitation to act as a cartel by bidding less on the licenses. On the other hand, a winner's curse phenomenon may have *really* been hidden below the data. How can we decide? The problem is that field data cannot settle the dispute because they do not provide information concerning crucial variables such as agents' private valuations or the real profitability of an oil lease in the long run. John Kagel and Dan Levin (1986) tried to tackle the problem by reproducing the winner's curse phenomenon experimentally. In the laboratory, they provided their bidders with information about the possible value of an auctioned item by privately communicating to each subject a value x_i drawn from a uniform distribution $[x_0 - \varepsilon, x_0 + \varepsilon]$, where x_0 is the real value of the item (i.e., the sum experimenters will pay the winner) randomly drawn from a uniform distribution on an interval $[x^*, x^{**}]$. The experimenters communicated to the bidders the range of ε and computed for them the upper and lower bound for the value of x_0 ($min\{x_i + \varepsilon, x^{**}\}$ and $max\{x_i - \varepsilon, x^*\}$, respectively).

The idea of modeling bidders' uncertainty as a random draw from a uniform distribution comes from auction theory, more precisely from the "common value model" first devised by Wilson (1977) and later refined by Milgrom and Weber (1982). The model is based on four fundamental assumptions: (1) that values are common and unknown to all, (2) that bidders are symmetric, (3) that the payoff is a function of bids alone, and (4) that bidders are risk neutral. The first three assumptions seem to be empirically justified in the OCS case, whereas risk neutrality is needed for analytical reasons (assuming risk aversion, for example, leads to ambiguous results in a number of cases) (cf. Milgrom 1989; McAfee and McMillan 1987). The solution to the common value model is known as "noncooperative equilibrium with risk-neutral bidders" (or RNNE for short) and predicts that the agent with the highest private signal (denoted by x_1) will generally win the auction. However, if bidders are rational maximizers as the RNNE models assumes, the one with the highest signal is supposed to revise her valuation of the item in light of the fact that her private information signal is likely to be *too* high. In technical terms, this revision is required to avoid an "adverse selection problem." Denoting the expected value conditional on having the highest information signal as $E[x_0 \mid X_i = x_1]$, the winner is cursed every time the actual estimate of value exceeds the latter, that is, whenever

$$(\text{WC})\, E[x_0 \mid X_i] > E[x_0 \mid X_i = x_i].$$

In this case, in fact, the winner fails to take into account the adverse selection problem and therefore will experience, on average, negative profits. (Notice that the inequality (WC) above is best characterized as a *hypothesis* rather than as a proper theory: unlike the RNNE solution, it does not follow from a formal model nor does it provide a full explanation of the bidding process. It is defined as a contrast case, from the hypothesis that real bidders are not fully rational and fail to revise their expected values correctly.) If the RNNE model is right and bidders really are rational maximizers, the winner's curse should not occur, and the evidence presented by the Atlantic Richfield Company should be explained in a different way.

The goal of the experiments devised by Kagel and Levin was to show how certain features of the OCS data can be reproduced in the laboratory. The winner's curse hypothesis was tested by controlling for the number of subjects and the nature of the information (public vs. private).

Number of subjects. Kagel and Levin ran experiments with a "large" number of bidders and experiments with a "small" number. When the number of competitors grows, a rational agent is supposed to take into account two opposite considerations: one should bid more aggressively because the signal values are more congested, but less aggressively because the adverse selection problem becomes more severe. In fact, an RNNE bid function taking into account these considerations requires the bids to remain constant or to decrease as the number of competitors grows.[6] If the winner's curse explanation is right, in contrast, higher bids should be observed as the number of competitors increases. Varying the number of bidders thus provides a means for discriminating between the two rival hypotheses.

Information. Some experiments involved only private information signals, whereas others involved public information: bidders were asked to provide a first evaluation under knowledge of x_i only, and then a second one after having been given some additional public information signal x_p (the lowest of the private signals formerly distributed, x_L, turns out to be particularly convenient for analytical purposes). The public information control provides useful insights into the bidding mechanism. In RNNE, in fact, public information is supposed to raise the bids of all the subjects who have not received the highest private signal; this should put pressure on the x_1 bidder (the winner, according to RNNE) and hence reduce her profits by almost one half.[7]

---

[6] See Kagel and Levin (1986) for the quantitative analysis behind such a prediction.

[7] From $E[\Pi \mid W] = 2\varepsilon/(N+1) - Y$ (where N is the number of bidders in the auction and Y is a negative exponential that becomes rapidly negligible as the value of x_i departs from extremely low values) to $E[\Pi \mid W, X_L] = \varepsilon/(N+1)$. The details of such a prediction can be found in Kagel and Levin (1986), but some amendments are due in light of Cox, Dinkin, and Smith (1999) and Campbell, Kagel, and Levin (1999).

Kagel and Levin (1986) observed two results. (1) In "small group" experiments, the winners bought the items at a profitable price, but the profits were considerably lower than those predicted by the RNNE model (65.1 percent of the latter). In "large group" experiments, the winners experienced losses on average. (2) In auctions with a small number of bidders, the injection of public information raised prices; when the number of bidders was large, in contrast, prices fell, contrary to the RNNE prediction. Both results are consistent with a winner's curse explanation. Winners, ex hypothesis, overestimate values; public information tends to reduce uncertainty about the true value of the item, so that bidders with the highest private information can revise their valuations.

Tightening the bridge

In their 1986 paper, Kagel and Levin claim they have produced the winner's curse phenomenon in the laboratory. Initially, two alternative explanations of the data were available and the experiment was designed to test them. However, the experimental result is still confined to laboratory conditions. The winner's curse can be produced experimentally, but does it exist in the wild? In order to answer this question, Kagel and Levin focus on some similarities between a laboratory and a real-world phenomenon first observed by Mead, Moseidjord, and Sorensen (1983). Mead and his colleagues had collected data about the different levels of profits achieved by oil companies in so-called wildcat and in "drainage" leases. Wildcat leases are more risky because no evidence about the past productivity of these isolated tracts is available. In contrast, drainage leases are on tracts lying adjacent to some hydrocarbon reservoir. The developers of the adjacent tract (the "neighbors") have higher private information on the profitability of the drainage tract, but all bidders ("non-neighbors") know that something is likely to be found.

Mead et al. (1983) noticed that in the Gulf of Mexico from 1954 to 1969, both neighbors and nonneighbors had on average higher rates of returns in drainage than in wildcat leases, a fact that is incompatible with the RNNE explanation. Why? In RNNE, depending on whether the information available is (a) purely public, (b) purely private, or (c) both private and public, we should expect rates of return (a) lower for all or (b) and (c) higher for neighbors than nonneighbors, with the latter earning less than they would in absence of insider information. Intuitively, with perfect rationality, public information should allow a more accurate estimate of the real value of the tract, thus making competition stronger and depressing the winner's profits. If a winner's curse effect is present, in contrast, the observed phenomenon can be easily explained: an increase in insiders' information reduces the winner's overestimation of a tract and thus raises the returns of both neighbors and nonneighbors. These field data match Kagel and Levin's experimental results remarkably well. In fact, Kagel and Levin managed to replicate these data in the laboratory, where one can control for public information at will.

More generally, Kagel and Levin's procedure can be analyzed as follows. Let us call the evidence in need of explanation, that is, the fact that oil companies in the Gulf of Mexico experience on average low returns from their leases, e. The goal of the experiment is to discriminate between two alternative theoretical hypotheses H_1 and H_2 – the RNNE model and the winner's curse hypothesis, respectively. The construction of an experimental common value auction allows one to test new predictions from H_1 and H_2. Kagel and Levin, by varying initial conditions such as public/private information and the number of bidders, construct a test that is moreover a *quasi*-crucial experiment with respect to H_1 and H_2, that is, such that H_2 & $K_i \rightarrow e'$ but H_1 & $K_i \rightarrow \sim e'$. The new evidence e' collected in the laboratory confirms that a winner's curse phenomenon is likely to be hidden behind experimental bidding. The experimenters, however, are aware that such evidence (e') cannot settle the dispute about the target system. Therefore, they show that in the real world, there are cases of variation of public/private information analogous to those reproduced in the laboratory. In the OCS case, the field evidence was provided by the study of Mead and his colleagues. The argument for external validity consists of showing that there are analogies between experimental and real-world phenomena.[8]

The first moral is that experimental evidence can help only at an intermediate stage of confirmation. It cannot completely bridge the gap

---

[8] Notice incidentally that the correspondence between experimental and field evidence is, strictly speaking, of the "phenomena to phenomena" kind, to use Bogen and Woodward's (1988) terminology (see also Chapter 4). The calculation of profits is rather straightforward in the experiment: it is the difference between the value of the auctioned item (x_0) and the price paid for it (b_1) (this is one reason one does experiments in the first place!). In the field, things are more complicated; Mead et al. (1983) use the so-called internal rate of return (IRR) measure: the rate of discount that makes the present value of the stream of net revenue (i.e., gross revenue minus costs minus taxes) equal to zero. The IRR can only be estimated because there do not exist direct data for a number of costs (e.g., exploration costs, postsale exploration, drilling, development, production, interests, and abandonment costs) that must be derived from other indicators, and the taxes attributable to each lease can only be calculated on the basis of estimated costs and revenues.

between the real-world phenomenon and the hypothesis under test. Experiments can increase the plausibility of an explanation, but only up to a certain point. The reason is not only that a pattern of data can be explained by different theories, but that it may also result from a different causal process. A Duhem-Quine problem can be tackled in the laboratory by controlled testing, but establishing that a certain explanation is the right one in the (artificial) domain X does not prove that the same process lies at the origins of a similar pattern of data in the target domain Y. In order to be convinced that this is the case, one needs some further independent evidence from the target domain of application – the real-world phenomenon one is interested in understanding in the first place.

Analogical reasoning

What kind of lesson do the FCC and OCS cases teach us? Is it possible to sketch a general account of external validity inferences? Such an account will have to be at least compatible with the theory of internal validity outlined in the first part of the book; but, in fact, we can be more ambitious: it will turn out that both inferences are instances of the eliminative inductive method. Recall the basic idea defended in Chapter 7: experimenters proceed by eliminating potential flaws in the experiment (or artifacts), particularly by making sure that no uncontrolled factor confounds the inference from the evidence to the main hypothesis under test.

Now, what sort of thing can lead one into error while extending a laboratory result to the field? In making external validity inferences, one can make a number of different mistakes. One may observe phenomenon Y in the lab and incorrectly infer that the same phenomenon also takes place (or can take place) in a given field setting. Or one may establish that X causes Y in experiment E but erroneously infer that X also causes Y under nonlaboratory circumstances F, G, and so on. First of all, consider that in order to identify a flaw in an external validity inference, one must specify a real-world target. If you worry that Y may not occur out of the lab *generically*, there is little you can do to figure it out. Instead, scientists usually worry about the extension of an experimental result to a specific target: to a population of patients who suffer from a certain disease, for example, or to a market with specific characteristics. The obvious thing to do, then, is to go out and have a look: if you observe the right sequence of Xs and Ys in the target, for instance, you will be encouraged to believe that because X causes Y in the lab, it does the same in the target. The temptation to frame this approach in terms of "analogical inference" is strong. In fact, the idea that external validity inferences are analogical in character has been recently defended by Hugh LaFollette and Niall Shanks (1995), Paul Thagard (1999), and myself (Guala 1998).

Analogy originates from the Greek word *analogia*, 'according to a ratio.' In the Pythagorean tradition, more precisely, an analogy was an identity of ratios. This meaning survived in the mathematical sense of analogy as proportion: $a : b = c : d$. Whereas according to the everyday meaning of the term, an analogy involves two entities; in the original sense, an analogy always involves at least four terms taken in couples: "As A is to B, so C is to D," according to Aristotle (*Topics*, i, 17). Analogies have a well-known heuristic value: by postulating an analogy between two sets of properties, we can infer the existence of a hidden property in one set from the observation of some properties in the other set. Analogical models sometimes work precisely this way: by observing the properties of a model we are induced to think that similar properties are to be found in the real entity modeled. (Famous examples in the history of science include Watson and Crick's "staircase" or double-helix model of DNA, and the model of the atom as a small solar system; cf. Giere 1979, Ch. 2.)

Experimental systems sometimes play a similar heuristic role. In the early fifties, experiments on cynomolgus monkeys helped enormously in the discovery of the mechanism of propagation of poliomyelitis. In the laboratory, experimenters knew exactly when a monkey had been fed (i.e., when it had contracted the poliovirus) and could determine precisely when the virus appeared in its blood. By means of analogical inference, the scientists conjectured that the period of incubation must be approximately the same in human beings and, in fact, found the virus in the blood of patients who were in approximately the same stage of illness (cf. Paul 1971, Ch. 36). To say it once again with Aristotle, "A is in B like C is in D." More rigorously, in mathematics, the knowledge of three terms of a postulated proportion such as $1 : 3 = x : 6$ leads to knowledge of the value of $x = 2$.

However, in mathematical examples, the analogical relation has been *postulated*. In empirical contexts like those we are interested in, the relation is an *empirical hypothesis*. The empirical hypothesis is used to make inferences, including external validity inferences. In biomedical research, for example, the inference may take this form (adapted from Thagard 1999, p. 140):

(1) Humans have symptoms (disease) *Y*.
(2) Laboratory animals have symptoms (disease) *Y*.
(3) In laboratory animals, the symptoms (disease) are caused by factor
(virus, bacteria, toxin, deficiency) *X*.
(4) The human disease is therefore also caused by *X*.

Analogical reasoning is fallible and, in fact, sometimes leads to error.[9] However, it would be wrong to criticize analogical reasoning simply because it *may* lead to a mistake. All inductive inferences are fallible (otherwise, they would be deductive rather than inductive in the first place), and external validity inferences surely involve an inductive step of some sort. The interesting point is, rather, How do we distinguish reliable from unreliable inferences? Drawing analogies is a mapping procedure, in which elements of a set of properties of an object are put in correspondence with elements or properties from another set or object. However, every object is similar to every other object in an infinite number of respects. There are potentially an infinite number of maps to be drawn, many of which will be uninteresting or even misleading from a scientific viewpoint.

Strengthening the analogy

To put it another way, consider the role played by statistical regularities in *internal* validity inferences. Under the right circumstances – for example, in the context of a well-designed experiment – statistical associations may constitute evidence for causation. However, the very same correlations observed in uncontrolled circumstances do not bear equal weight. One thing is to observe that factors *X* and *Y* are regularly associated in the field, where many (uncontrolled) factors could have been responsible for their instantiation. Quite another is to "trigger" *X* and observe *Y* in the context of an accurately designed experiment, in which the "other

---

[9] Take the case of poliomyelitis once again. In order to investigate its mechanism of propagation, Simon Flexner and Paul Lewis (1910) studied the process of infection in rhesus monkeys. Apparently, monkeys are easily infected via the nose, from which the polio virus travels to the olfactory nerves and finally to the spinal cord. Nasal sprays based on alum, zinc sulfate, and picric acid seemed to be able to kill the virus and were therefore tested in experiments on humans. However, the human nervous system is less susceptible to poliomyelitis than that of lower primates; in contrast, our intestinal tract is weaker and is attacked easily by the poliovirus, whereas monkeys are more resistant to this kind of infection. Two different causal mechanisms in this case led to the same effects, and the analogy proved to be misleading. (Cf. LaFollette and Shanks 1995, pp. 126–8; for a more detailed account of this episode, see Paul 1971, Chs. 12–23.)

factors" have been appropriately shielded or controlled for. A correlation between X and Y provides evidence for the hypothesis that X causes Y only if we have made sure (by means of appropriate experimental design and data analysis) that the correlation could only emerge from *that* data-generating process. The circumstances matter: the same data can bear different weight depending on whether the background circumstances are "right" or not. The same moral applies to external validity inferences: analogical correspondences are not enough. We need "strong" analogies – but what makes an analogy strong in the first place?

The goal, which should be familiar by now, is to create (or select) circumstances under which it is unlikely to observe evidence of a certain kind unless the external validity hypothesis is true. In this case, the evidence is the correspondence between observed features of the target and observed features of the experimental system; the external validity hypothesis is that the relata belong to similar causal mechanisms. The probability of observing such a correspondence (were the hypothesis false) is low if we have eliminated alternative reasons why such a correspondence might occur, other than the causal similarity between the two systems. If you want to generalize from A to B, you should make sure that A and B are as similar as possible.

Remember that external validity inferences are inferences to circumstances that we *know* to be different in some respects from the experimental situation. In order to make such inferences reliably, we must ask (and check) whether the differences between the experimental and the target system can confound the external validity inference or not. You cannot extend to human beings the results of experiments on mice, for example, unless you have good (experimental) grounds to believe that certain differences between the anatomies of mice and human beings do not matter (i.e., they are not error-generating differences).

This is, in fact, the logic underlying the best-known external validity control – representative sampling. If you want to generalize to population B, you should make sure that you have in the lab good representatives of the individuals in B (you need students if you want to generalize to students, housewives for housewives, mammals for mammals, etc.). Unfortunately, in most cases, this is just the beginning of the story: experimental conditions include not only a pool of subjects, but also a range of environmental factors, treatments, and boundary conditions in general. In many cases, it is the environment and the treatment that worry us the most. (Think of the stylized, abstract tasks of experimental cognitive psychology and decision theory, for example.) For this reason, experimenters try to make sure that as many differences as possible between experiment and target are checked in the laboratory by incorporating them one by one in the experimental design. The logic of inductive inference is *eliminative*, both *within* and *from* the experiment.[10]

The FCC case displays such a strategy in full-fledged form because the experimenters made sure that the real auctions (the target) were modeled on the experimental prototypes. However, in other cases, one can work backward from target to experiment to achieve the same result. The standard sequence of trials to test drugs in experimental medicine is a good example of a compromise: experimenters start with animals,[11] move on to human beings in "ideal" experimental settings, and conclude with so-called efficacy trials with patients in more realistic conditions (which does not mean that real-world conditions themselves cannot sometimes be modified to make the drug more efficacious or to avoid unpleasant side effects – that's what hospitals are for, among other things).

Whereas the FCC project started with an ideal mechanism of allocation to be created from scratch, in the winner's curse case, experimenters tried to mimic an already existent market. They worked, in other words, with a real-world target in mind. The essential steps in the procedure are represented in Figure 9.1. The first step is one of modeling the fundamental features of the OCS auctions, something that had already been achieved by the game theoretic literature of the seventies and early eighties. Some of the assumptions of these models are controversial, however. In particular, the controversy in this case revolves around the perfect rationality imputed to bidders and their capacity to solve the adverse selection problem. Two different hypotheses can be derived from the auction models depending on whether we use, respectively, the perfect rationality assumption (RNNE hypothesis) or the imperfect rationality assumption (winner's curse hypothesis).

The uncontroversial assumptions can be used to construct an experimental system that is not biased against any one rival hypothesis. (I call them "neutral" assumptions in Figure 9.1.) These include, for instance,

---

[10] This does not mean, of course, that the elimination of all possible sources of error can be always carried out in full. Practical or ethical problems may, in fact, prevent the experimenter from testing certain hypotheses, and in some cases, it could be argued that the experiment presents some noneliminable differences from its target that prevent it in principle from being a useful replica of the real world (Bardsley, in press, examines such a possibility).

[11] I am simplifying drastically here: to find out which animals are "right" for which kind of investigation is not a trivial matter. On animal models in biomedical science, see e.g., Kohler (1994), LaFollette and Shanks (1995), and Ankeny (2001).

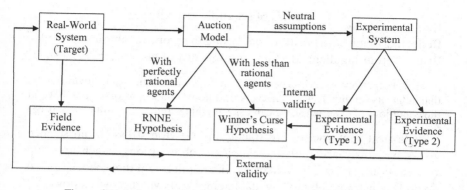

Figure 9.1. Inferences *in* and *from* the experiment: the OCS case.

that the value of the items to be auctioned is the same for all bidders but unknown to them, that the auction is run with a sealed bid system, that bidders are symmetric, and so on. It is crucial that these assumptions also mirror accurately the features of the real-world target (the OCS auctions). The experiment, in fact, generates two kinds of evidence. Evidence of type 1 is used to discriminate between the two rival hypotheses; evidence of type 2 is used to bridge the gap between the laboratory and the real world by drawing an analogy with already existing field evidence. However, the analogy cannot be strong unless experimental and field evidence have been generated by systems that are similar in all relevant respects or, in other words, unless all sources of external validity error have been taken care of by means of accurate design. Strong external validity inferences begin and end in the field.

Notice the role played by theoretical models. Contrary to the theory-testing view, they are neither the beginning nor the end of the story. The aim of the experiment is not to test the models, which instead work as intermediate devices in the design of a good experiment – an experiment that says something useful about the real world rather than about the theory. When the model makes plausible assumptions, they are incorporated into the experimental design. When the assumptions are dubious (e.g., the assumptions about bidders' cognitive capacities), the experiment can test them empirically. Models and experiments complement each other and constitute intermediate steps on the route toward the understanding of a real-world economic system.[12]

---

[12] Notice that the *combined* use of field and laboratory evidence allows one to avoid an apparent "regress": if experimental results require field evidence in order to solve external validity problems, why should we do experiments in the first place (Siakantaris 2000)?

Models of scientific growth

Reflecting on his own position concerning the scope of laboratory results, Bruno Latour highlights the lack of originality of his "radical localism":

A scientific fact never survives beyond the network of practices and circumstances that define its validity. This is [. . .] a very traditional philosophy of science: beyond the conditions of experience, we can't say anything credible. (Latour 1988, p. 68)

"Classical" empiricists like Bacon and Newton stressed that science grows by slowly accumulating empirical observations, which are only very cautiously turned into theories. The requirement that speculation be severely constrained by the available empirical evidence was a powerful rhetorical weapon against Aristotelian defenses based on metaphysical speculation or mere appeal to authority. The classical empiricists portrayed scientists as a group of boring collectors of evidence who refuse to generalize until enough data have been accumulated. So perhaps Latour was defending good old classical empiricism all along.

The classical empiricist view has been criticized and sometimes ridiculed as "naïve inductivism" (or worse). Hypothetico-deductivists have dropped the requirement that theories should stick as close to the facts as possible. Popper challenged common sense even further by claiming that bolder theories that stretch considerably beyond the evidence are more desirable and "scientific" than those that do not. Other hypothetico-deductivists preferred to remain silent on the looseness that is allowed between evidence and hypothesis, presumably because they believed that such issue could be resolved only after an appropriate theory of inductive inference had been produced. *How much* beyond the evidence should a good theory or hypothesis be allowed to stretch? The examples discussed so far suggest that modesty is a virtue in experimental and applied science. Experiments, to begin with, are usually aimed at testing local hypotheses rather than grand theories. General theories and models are often used to suggest or derive such hypotheses but receive little reward back from experimental investigation. Experimental evidence tends to confirm or refute only local hypotheses, or small portions of general theories. Eddington's eclipse observations, to take a famous example, only proved that light bends in proximity to objects with a big mass. They did not, and

---

The answer is that the two types of evidence complement each other: what you need to know to use field evidence effectively in external validity inferences (*after* the experiment) is not what you needed to know to use it effectively *before* the experiment was run. A good external validity inference always builds upon existing experimental knowledge to learn something new (see also Guala 2002b).

could not, provide a confirmation of Einstein's General Relativity *as a whole* because several other nonrelativistic gravitational theories imply approximately the same angle of deflection as Einstein's.[13] Moreover, as highlighted in Chapter 7, experiments usually generate knowledge of peculiar portions of reality – of artificial systems that have been especially created by experimenters in order to answer specific questions. Given the trade-off between internal and external validity, experimental results are more reliable when the artificiality is greater – hence the difficulty in extending such results to other circumstances outside the laboratory.

Of course we infer all the time from what happens under one set of experimental circumstances (under which a hypothesis has been tested) to what is likely to happen in other, similar circumstances. What counts as "similar" in such cases is usually determined by theoretical insight of some kind. We obviously do that because we cannot test a given hypothesis in *all* possible circumstances. However, it is one thing to use theory to project beyond the conditions of experience and to formulate testable predictions during the heuristic stages of a research program; it is quite another to *act* on the basis of untested conjectures. When it comes to application, scientists want to carefully test every aspect of a design or a technology. The experiments on the FCC auctions are a good example in this respect. However, other cases of science with high stakes can be easily found, such as the launch of a space shuttle or the commercialization of a new drug. The instructions for most medicines in your cabinet, for example, say that they should not be taken by pregnant women except under a doctor's supervision or in exceptional cases. This usually happens not because it is known that the drug is harmful to pregnant women, but because we lack experimental data on women in that particular condition. When science matters, there is no substitute for careful empirical checking.

The examples discussed in the last two chapters suggest that when scientists try to extend the scope of their experimental results, they move cautiously, by gradually modifying the target system, the experimental system, or both, so they resemble each other, and by checking that the remaining dissimilarities do not matter. The growth of scientific knowledge, then, seems to proceed by careful induction from the particular to the particular, rather than by means of bold theoretical generalizations. Theories do play a role in the design of experiments and at various stages

---

[13] Of course, a long series of experiments could, in principle, test a "grand" theory bit by bit, although the historical record suggests that the controversies surrounding grand theories are rarely resolved this way. See Mayo (1996, Ch. 8) for further discussion.

in the process of inductive inference, but at both the starting and the end points of the inference, we find knowledge of rather specific systems. This vindicates, in part, the classical empiricists, with their "boring" picture of scientists slowly accumulating empirical facts. Perhaps some parts of science are driven by abstract theoretical speculation, with little or no empirical input (it is easy to find such cases, from Superstring theory in physics to General Equilibrium analysis in economics). However, when science is applied and when it is important that the applications work reliably, we invariably find that bold science is abandoned in favor of boring empirical checking.

TEN

Experiments as Mediators

Two theses are prominent in the second part of this book. The first one is that we have no reason to believe a priori that an experimental result applies (or does not apply) to nonexperimental circumstances. The second thesis then follows quite naturally: successful external validity arguments are empirical and can be constructed only by appropriately combining experimental and field evidence. I have also tried to specify what counts as "appropriate" in this context, or what sort of requirements a strong external validity inference should satisfy.

In a way, at this stage I consider the important job to be done, but not because everything has been said on how experimental and field data can be used to draw external validity inferences. On the contrary, the analysis in Chapters 8 and 9, being so tightly linked to concrete example, is almost certainly incomplete. However, having discussed two paradigmatic examples and having sketched a broad theory of inference based on eliminative induction, I consider the definition of more sophisticated strategies an exciting research agenda for the future. In what remains of this book, then, I would like to elaborate on the image of experimental economics that emerges from the discussion so far.

It is important, for example, to realize that by endorsing the above two theses, one subscribes to a very specific view of experiments and their role in scientific discovery. This chapter is devoted to articulating such a point of view and to comparing it with alternative accounts of the role of experimental science. I argue that experiments are "intermediate" steps in the long path leading from the formulation of ideas or hypotheses about the real world to their final application. As such, they share some characteristics that are usually attributed to theoretical models and simulations.

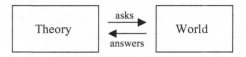

Figure 10.1. The naïve view.

Obviously, however, they also differ from models and simulations in some crucial ways. The task, then, is to figure out exactly where the analogy ends and the disanalogies begin. I also argue that experiments come in different degrees of concreteness, from applied experiments devoted to testing external validity claims to "middle-range" experiments devoted to testing the robustness of economic phenomena or results, to abstract experiments devoted to testing highly simplified theoretical models. As we shall see, the distinction is far from trivial and is often overlooked in experimental economics debates. By recognizing the difference between types of experiments, we can dispel a good deal of unnecessary confusion about the way in which experimental results are (and should be) generalized to other circumstances of interest.

A naïve account

Let me start from an admittedly oversimplified account of empirical testing. According to this view, science is a game with two players (Figure 10.1). On the one hand, we have theory, on the other, empirical reality. Theory is used to ask questions, the external world answers these questions, and the theory is successively modified in light of the answers. By iterating such a procedure, scientists devise better and better representations of their subject matter.

This account, however, fails to capture the complex process leading from abstract theorizing to application. In trying to articulate such a process, I am concerned in particular with the place of experiments. Where are they located? At what stage do they play a role in scientific discovery and testing? In the naïve account, experiments seem to be quite naturally located on the right-hand side of the picture. Indeed, traditional empiricist philosophies tend to draw a sharp distinction between descriptive or representational devices and what is described or represented. Theories, models, and simulations are customarily placed among the representational tools, whereas experiments are seen as parts of the natural or social world that have been carefully designed in order to test the representations. I argue that, in contrast, it is useful to think of models and laboratory experiments in economics as tools of the same kind, somehow located between our statements about the "real world" (call them scientific laws, principles, theories, axioms, hypotheses) and the world itself. Borrowing from Margaret Morrison and Mary Morgan (1999), I say that such entities "mediate" between theory and reality. The rest of this chapter is devoted to clarifying this vague proposition and illustrating it by means of examples.

Theoretical models

First of all, it is necessary to sort out my terminology. Having distinguished between theoretical and other (experimental, low-level) hypotheses in Chapter 3, I have nevertheless spoken very loosely and freely of "theories" and "models," as if they were the same kind of thing. In fact, it is worth distinguishing between them: the basic unit of theorizing in economics is the *model*, and economic *theories* are *sets of models*. Modeling has become such an essential element of their practice that economists tend to forget how recent an acquisition it is. Before the 1950s, economists made occasional use of models and used to theorize in a variety of ways, but with the postwar mathematization of economic theory, modeling has become so central that nowadays you cannot really claim to be an economist unless you know how to build a model. The term *theory* has not disappeared but has become a loose term to define broad categories of theoretical models inspired by some common general principle or domain of application (as in the expression *game theory*, *theory of the firm*, or *rational expectations theory*). A theory is a set of models with some family resemblance, to use a Wittgensteinian expression.

What is a model then? As usual, it is best to answer with a concrete example in mind. I follow a recent analysis by Robert Sugden (2000) and focus on Thomas Schelling's famous checkerboard model in *Micromotives and Macrobehavior* (1978). Schelling's goal is to give an account of racial segregation in American cities, but the goal is achieved by describing an imaginary checkerboard with a particular tessellation, upon which coins of two kinds (dimes and pennies) move according to precise rules. Schelling shows that when some specific conditions hold, particularly when certain "preferences" about the occupants of each coin's neighbor squares hold, then certain arrangements of the coins on the checkerboard will follow. Informally (but rigorous rules of the game can be provided), each coin moves in an attempt to escape from areas where an overwhelming majority of coins of the other type prevails (say, two thirds or more). Every time a coin is surrounded by a majority of another type, it is moved. Via successive reshuffling, it is shown that a complete separation of dimes and pennies is produced on the checkerboard.

Schelling's paper is an exercise in analyzing the dynamics of an abstract toy-model, but the story about the model-world is intended to support a hypothetical explanation of the evolution of racial segregation in the real world.[1] Sugden argues that "moving from the model to the hypothesis requires a step in the argument which most readers would be willing to make, but for which no formal justification was available." According to Sugden, such a step is an inductive one.

What Schelling has done is to construct a set of *imaginary* cities, whose working we can easily understand. In these cities, racial segregation evolves only if people have preferences about the racial mix of their neighbours, but strong segregation evolves even if those preferences are quite mild. In these imaginary cities, we also find that the spatial boundaries between the races tend to move over time, while segregation is preserved. We are invited to make the inductive inference that similar causal processes apply in real multi-ethnic cities. We now look at such cities. Here too we find strong spatial segregation between ethnic groups, and here too we find that the boundaries between groups move over time. Since the same effects are found in both real and imaginary cities, it is at least credible to suppose that the same causes are responsible. Thus, we have been given some reason to think that segregation in real cities is caused by preferences for segregation, and that the extent of segregation is invariant to changes in the strength of such preferences. (Sugden 2000, p. 24)

Thus, the activity of modeling in economics has to do with the construction, and the theoretical description, of "model-worlds." Such systems are usually abstract entities, existing only in the minds of those who happen to read, for instance, Schelling's book. However, in principle, a real, material checkerboard could be manufactured with its dimes and pennies and a "segregation game" played for real (Schelling actually invites his readers to do so). Schelling's game theoretic account of segregation is *true* of the checkerboard described in his book – and trivially so, because the checkerboard system is designed to perfectly satisfy Schelling's game theoretic rules. Schelling's theory of segregation is a nice example because the model here (the checkerboard, the dimes and pennies, and the rules

---

[1] The "dictionary" translating the model into a real-world representation is as follows: coins = people, dimes and pennies = two races, areas = neighborhoods, separation = racial segregation, rules = people's preferences, etc. As Dan Hausman has pointed out (in conversation), Schelling's contribution can also be read as the exploration of an interesting and counterintuitive *possibility* rather than as an explanation of real-world behavior. Sugden (2000) discusses and dismisses this interpretation.

of the game) can be easily identified and separated from the theoretical description. However, the same applies to all theories: provided their axioms and principles are consistent, an abstract model can be identified of which the formalism is true – or which is "picked up" by the formalism, so to speak.

The model-formalism distinction should be easy to grasp for those trained in neoclassical economics. Take, for example, the Walrasian auctioneer that is central in general equilibrium theory. It is a typical model in the sense above: it is an abstract entity because no real market uses *tâtonnement* to determine prices (although a few market institutions are similar to the Walrasian auctioneer). And it is an entity of which the theory's equations are true: if such an institution existed, then Walrasian equilibrium theory would fit it perfectly. Indeed, Walras in the fourth edition of the *Elements of Pure Economics* seems to suggest that the term *tâtonnement* refers to the technique of solving a system of simultaneous equations by iteration.[2] This ambiguity (*tâtonnement* as what the auctioneer does, or as what the theorist does?) just highlights the above point about the nature of models: they are entities (abstract or concrete) whose precise properties are defined by the theoretical formalism.

A theory's formalism, then (i.e., the set of "axioms," "principles," or "laws" of the theory), is not applied directly to reality but is first and foremost asserted to be true of an ideal model, *then* is suggested to be somehow relevant for the understanding of some real-world phenomena. The path from theoretical speculation to the real world is broken down into at least two steps. Indeed, some philosophers of science have proposed to take the models of which theoretical principles are true as the primary unit that defines what a scientific theory is. Such an idea – defended, among others, by Bas van Fraassen (1980), Ronald Giere (1988), and Frederick Suppe (1989) – is usually referred to as the "semantic view" of theories.

The semantic view is more a family of doctrines than a single, unified philosophical theory, but all its versions share a distaste for the older approach (the so-called standard view of theories that I briefly discuss in Chapter 3), according to which theories are basically sets of *statements* (laws and bridge principles). In the semantic approach, the fundamental component of a theory, the model, is in contrast a *structure* – an entity or set of entities (a system) with specific properties and relations among them and/or their parts – that satisfies the linguistic elements of the theory. The

[2] On the Walrasian auctioneer and its various possible interpretations, see de Vroey (1998).

latter are secondary in the sense that they can be formulated in various equivalent ways, as long as they are satisfied by the models that make up the theory. The precise form of the axioms, laws, and so on, may change depending on the language and axiomatic system that scientists choose, but the models will remain the same.

Besides preserving the identity of the theory in the face of changes in linguistic formulation or axiomatization, the semantic view has other advantages. First, by being neutral on the "materiality" of models, it can be stretched to cover models of very different kinds – concrete physical objects and mathematical structures, as well as the abstract or mental models that are the subject of so-called thought experiments.[3] Moreover, the semantic view avoids some puzzles concerning the applicability of models to the real world. Most scientific theories do not describe anything that exists, or can even possibly exist, in reality. Newtonian mechanics is true of dimensionless mass points, general equilibrium theory is true of frictionless economies populated by perfectly rational and omniscient decision makers, and the list of examples could continue to include the most interesting and empirically successful scientific theories. However, it is more reasonable to say that the statements (laws, axioms, principles) of a theory are not even *supposed* to be true of anything real. They describe idealized or fictional entities, which are then used to understand what goes on in the real world.

As with most philosophical explications of scientific concepts, one can ask whether the semantic view is an accurate description of the way in which scientists speak of models and theories. There is no doubt that the semantic view captures the spirit of many remarks made by economists. However, sometimes economists also speak of models as if they were linguistic entities (e.g., when they refer to the axioms of expected utility theory as constituting – rather than defining or describing – the model of a perfectly rational agent).[4] For this reason, I do not want to make any claim of greater descriptive accuracy in subscribing to the semantic

---

[3] By taking this position, I'm implicitly denying the methodological specificity of thought experiments, which are assimilated to experiments (or "manipulations") of abstract models. There is no general agreement on this point, however, and the debate on thought experiments is nowadays quite lively in philosophy of science; see e.g., Brown (1991), Sorensen (1992), and Norton (1996).

[4] Thus Hausman (1992a, Ch. 5), e.g., following Giere's (1979) early work, argues that economic models are best interpreted as *definitions of predicates* (e.g., the axioms of expected utility theory define the predicate "is a rational economic agent"). Given that Giere has since changed his mind, Hausman seems to be the only supporter of the "predicate view" of theories.

view in the above form. To conceive of models as entities rather than statements makes a difference primarily in terms of conceptual clarity and ease of presentation. For my purposes, it also helps to highlight the analogies (and disanalogies) between models and experiments. The proof of the pudding is in the eating, and I leave it to the reader to decide at the end of the chapter whether the move was worth making in the first place.

Experiments as mediators

The route from what we say about the world and the world itself has already been broken into two substeps, with the models playing the roles of "mediators" in between. The idea of mediation has been used by Morrison and Morgan (1999) to capture a number of cognitive, practical, and pedagogical functions fulfilled by models. Here I would like to extend their account to suggest that models and experiments have a similar "mediating" role. Like models, *experiments in the nonlaboratory sciences mediate between what we say about the real world and the real world itself.* The notion of "nonlaboratory science" is derived from Ian Hacking:

Laboratory sciences are those whose claims to truth answer primarily to work done in the laboratory. They study phenomena that seldom or ever occur in a pure state before people have brought them under surveillance. Exaggerating a little, I say that the phenomena under study are created in the laboratory. (1992, p. 33)

The *non*laboratory sciences, then, are those whose claims to truth do *not* answer primarily to work done in the lab and that are aimed at studying phenomena that normally occur spontaneously outside laboratory walls. The laboratory in these disciplines is a tool, an instrument that scientists use in order to investigate nonlaboratory entities and phenomena.

In order to understand how experiments and models mediate, it is necessary to explicate the relationship among models, experiments, and the real world. Following the terminology already introduced informally, I call the real-world system (or set of systems) whose behavior we ultimately intend to investigate and understand, the *target system*, or simply the *target*. In sciences like economics, a target is typically a nonlaboratory entity, a naturally evolved economy that is too big and complicated to be fully controllable by economists. The study of the target, however, can sometimes proceed via the laboratory. Here's a typical route from theory to the real world: a model is used to give structure to a speculation

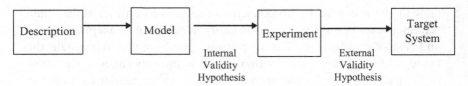

Figure 10.2. The route from theoretical speculation to the real world.

(a theoretical idea) about the economy. (To show, for example, that a certain phenomenon can occur in certain circumstances, what is the mechanism that can bring it about, etc.) Then, a specific hypothesis is generated from analyzing the model, for example by showing what would happen if . . . certain changes were made to a key variable. The hypothesis, however, is not tested directly on the target. The behavior of the target is likely to be too unruly to permit a valid test of the hypothesis.[5] Hence, a laboratory system is built (an experiment) that can provide an answer to the research question. Then, the experimental result is extended to the target by means of the external validity techniques discussed in Chapters 9 and 10. Figure 10.2 attempts to represent this process schematically.

This scheme is useful in order to highlight the mediating role of experiments, but should not be generalized too readily. I have stressed repeatedly that experiments do not always or commonly answer theoretical questions. Sometimes they replace models altogether, and sometimes they complement models, if the latter are too abstract or incomplete in some crucial respect. Figure 10.3 is an attempt to represent these two other very common cases.

In all cases, experiments require an *external validity hypothesis* stating that the laboratory system stands in some particular relationship to the target. (This is the main point, in fact, of using the "mediating" metaphor.) Experiments are used to "demonstrate" like models and simulations and somehow *stand for* a real-world target system, rather than *being* the target

---

[5] I use the term *hypothesis* here in a way that differs slightly from the standard terminology of the semantic view. A "theoretical hypothesis," according to semantic theorists, states that a model stands in a certain relation (of similarity, isomorphism, analogy, etc., depending on which version of the semantic view one subscribes to) with a set of real-world entities or systems. The nature of theoretical hypotheses in this sense is highly debated in philosophy of science (on the "stories" used by economists to connect models to their target systems, see also Morgan 2001). In taking the semantic view as my point of departure here, I primarily endorse the conception of models as systems and the idea that models are manipulated in order to see "what happens if" certain changes or interventions are made in highly controlled (indeed, in totally controlled) circumstances. The notion of representation is too grand a topic to be adequately discussed here.

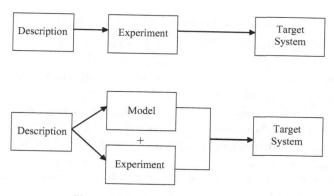

Figure 10.3. Two alternative routes.

system themselves.[6] They are just an intermediate step in the route from the realm of speculation to the real world. They are "mediators," in the sense that they help bridge the gap between our ideas and their intended domain of application. The naïve view encourages one to think of models (theories) and experiments as radically different things, whereas in reality they are not. Both models and experiments should be thought of as systems. Experimental systems are obviously more concrete than models and closer to their intended domain of application because they include features that are held in common with the target systems that we are eventually interested in understanding (the real-world economies). However, they are not the target system, and to move from experiment to target requires an inference. As I argued in Chapter 9, the experiment-to-target inference is – just like the model-to-experiment inference – inductive or ampliative (what we know about X does not enable one to derive deductively the properties of Y) and hence fallible.

Following the terminology introduced earlier, we may say that in the "laboratory sciences," experimenters "play" with the target system itself. In the "nonlaboratory sciences," it is sometimes possible to manipulate the target system in a nonlaboratory environment, but this is more often difficult, costly, dangerous, immoral, or even impossible, and the inferences drawn from uncontrolled experiments are hardly reliable anyway. Thus, *laboratory experiments in the nonlaboratory sciences demonstrate with experimental systems that "stand for" the target systems of interest.* This is my main claim concerning the nature and function of experiments in sciences such as economics.

---

[6] On models as "representatives," see Hughes (1999), and in the context of economics, Mäki (2001a).

Manipulating models and experiments

Of course, the claim has to be justified by showing that the mediators metaphor is fruitful in some interesting respects. The first important analogy between models and experiments is that both are manipulated in order to figure out "what happens if . . ." certain parts of the system are changed or varied systematically. This strategy, as we have seen, is absolutely central in laboratory experimentation, especially when scientists are testing causal hypotheses or claims; that it can be extended to model-based reasoning is perhaps more controversial. However, scientists are certainly not afraid to speak of the manipulation of models. Take, for instance, Robert Lucas:

> One of the functions of theoretical economics is to provide fully articulated, artificial economic systems that can serve as laboratories in which policies that would be prohibitively expensive to experiment with in actual economies can be tested out at much lower cost. (Lucas 1982, p. 271)

The key expressions in this quotation are *artificial economic systems* and *to experiment with*. First is the idea that models are things or systems, not statements. This may sound puzzling, given that you normally don't touch or see an economic model, whereas you have direct access to the axioms, principles, and assumptions of a theory. In fact, there are advantages to saying that the assumptions, principles, and so on *describe* the model, without being the model itself. As Lucas says, to propose a set of assumptions, postulates, and so on is a way of "providing" a model.

Then comes the most important idea, an idea that constitutes the core of the mediators approach (Morrison 1998b, Morrison and Morgan 1999): the model is an entity that can be "manipulated," or experimented upon, to put it in Lucas's terms. You *do things* with models, you don't just contemplate them or put them in correspondence with reality. I use the scare quotes because the manipulation of an abstract or fictional model is quite different from that of a concrete entity, of course. It is usually performed by making changes in the assumptions that describe some of the model's features and by demonstrating (perhaps mathematically) that certain interesting consequences follow from these changes. R. I. G. Hughes (1997) proposes the term *demonstration* to refer to model manipulation, regardless of whether the model is a concrete or an abstract entity. With models, whether abstract or material, we demonstrate: we trigger a mechanism and observe what it brings about. The mechanism may very well be purely logical (a set of rules of inference) and the consequence a theorem. In this sense, theorists do on paper something analogous to what experimenters do in the lab. When studying the physics of waves, for example,

[t]he same result appears whether we use the mathematical or the material model. The internal dynamic of the mathematical model is supplied by a mixture of geometry and algebra, that of the material model by the natural processes involved in the propagation of water waves. The internal dynamic of a computer simulation of the phenomenon would be something else again. But all these modes of representation share this common feature; they contain resources which enable us to demonstrate the results we are interested in. I choose the term "demonstration" in order to play upon its diachronic ambiguity. Whereas in the 17th century geometrical theorems were said to be "demonstrated," nowadays we demonstrate physical phenomena in the laboratory. Mathematical models enable us to demonstrate results in the first sense, material models in the second. (Hughes 1997, p. S332)

It is easy to realize that the notion of "demonstration" in this generic sense cuts across not only the distinction between concrete and abstract models, but also the one between models and experiments. In fact, I explicitly claim that models work in many ways like experiments. However, no matter how similar, models are not experiments.[7] In order to capture the crucial difference, it is necessary to introduce the notion of "simulation."

Models, simulations, and experiments

Nowadays, we are used to thinking of simulations as computationally intensive processes performed with the aid of powerful calculators. However, a simulation does not require a computer, in principle: a demonstration from a material model, for example, has all the essential characteristics of (and therefore can properly be called) a simulation.[8] What are these essential characteristics? Herbert Simon (1969, pp. 15–18) famously puts it as follows: simulations rely on a process of *abstraction* from the *fundamental principles* governing the behavior of the simulating and the target systems. If similar "organizational properties" arise at a certain nonfundamental level from different substrata, it is possible to abstract from the substrata and simulate the behavior of a system A by observing the behavior of another system B that happens to (or that is purposely built so as to) display those nonfundamental properties.

---

[7] On this specific point I disagree with the position outlined in Mäki (in press).

[8] The most famous example of a concrete simulating model in economics is the "Phillips machine." See Morgan and Boumans (2004) for a discussion and historical account.

Consider the ripple tank model again: at a general level of analysis, any kind of wave can be modeled as a perturbation in a medium determined by two forces: the external force producing the perturbation and the reacting force working to restore the medium at rest. General relationships such as Hooke's law or d'Alembert's equation hold for *all* kind of waves. More fundamental relationships, such as Maxwell's equations, describe the properties of the electric and magnetic fields only. The values given by Maxwell's equations can be used in d'Alembert's wave equation in order to obtain, for instance, the velocity of propagation of an electromagnetic wave because electricity behaves *like* a wave, although the fundamental principles are different from those at work, for example, in the case of water waves. The terms appearing in the equation describing the light wave and the water wave are to be interpreted differently in the two cases: the forces are different in nature and so are the two media in which waves travel. The similarity between the theoretical model of light waves and the ripple tank model holds at a very abstract level only. The two systems are made of different "stuff." Because of the formal similarity, though, the behavior of light waves can be *simulated* in a ripple tank. Both light waves and water waves obey the same nonstructural law, despite their being made of different stuff. This is because of different reasons in each case: different underlying processes produce similar behavior at an abstract level of analysis.[9] Similarly, human behavior can, to a certain extent, be simulated by means of computerized models but arises from "machines" made of flesh, blood, neurons, rather than silicon chips.

Working on this idea, we can devise a criterion to demarcate genuine experiments from "mere" simulations. The difference lies in the kind of relationship existing between, on the one hand, an *experiment* and its *target* system, and on the other, a *simulation* and its target. In the former case, the correspondence holds at a "deep," "material" level, whereas in the latter, the similarity is admittedly only abstract and formal. In a simulating device, the simulated properties, relations, or processes are generated by *different* (kinds of) *causes* altogether. In a genuine experiment, the same material causes as those in the target system are at work; in

---

[9] Of course, if one believes in the reductionist story according to which everything physical is made of the same fundamental subatomic particles, then both light and water waves are "made of the same stuff." However, the reductionist story is controversial (photons seem to have properties different from other particles), and at any rate, the fact that everything is made of the same stuff would not play any relevant role in explaining why both systems display certain nonfundamental relations.

a simulation, they are not, and the correspondence relation (of similarity or analogy) is purely formal in character.[10]

Considerations of this kind have sometimes been offered by economic experimenters to defend their methodology. Vernon Smith, for example, argues that "the laboratory becomes a place where real people earn real money for making real decisions about abstract claims that are just as 'real' as a share of General Motors." For this reason, "Laboratory experience suggests that all the characteristics of 'real world' behavior that we consider to be of primitive importance [. . .] arise naturally, indeed inevitably, in experimental settings" (1976, p. 274). This reasoning is used to support experimenters' confidence in their results. "Laboratory microeconomies are real live economic systems, which are certainly richer, behaviorally, than the systems parametrized in our theories" (1982, pp. 923–5). Experimental economies are supposed to work according to the same principles as the target systems in the intended domain of economic theory because the relevant components of the laboratory system are made of the same stuff.

Notice that these remarks should not be interpreted as a surrogate proof of the external validity of economic experiments. Such a proof, as we have seen and as Smith clearly recognizes, must eventually be empirical in character and can only be carried out by focusing on the *overall* causal isomorphism between an experiment and its target. It is not enough to show that properties X and Y are both present in the experimental and real-world systems and that they are connected in the "right" way. One also has to show that the other boundary or background conditions are arranged analogously in the two systems. Similarly, simulations also require some external validity inference before they can be applied to their targets. But the way in which this inference works is not quite the same as in the case of experiments, precisely because experiments and simulations are made of different stuff.

The most obvious situation in which experiments turn out to be useful is one in which some properties of the target system are, for some reason, obscure or contestable. If a model makes some disputable assumption about an aspect of the target system, it is sometimes possible to create an experiment reproducing that particular aspect in conditions that make

---

[10] See also Ernst Nagel's (1961, p. 110) distinction between "substantial" and "formal," or Mary Hesse's (1963) "material" and "formal" analogies (p. 63f). In Guala (2002a), I argue that some mediating entities share characteristics of both experiments and simulations. For the sake of simplicity, I ignore such "hybrids" here (but see also Morgan 2002).

its investigation easier. Of course, not all aspects of the target can (or should) be reproduced identically, for otherwise the main purpose of experimentation would be lost. The challenge is to reproduce *enough* of the target and in such a way that the substantial similarity between experiment and target is preserved. To simplify, suppose we are interested in the relation between two variables, X_1 and Y. It is important that we make sure that X_1 and Y are reproduced adequately in the experiment, but also that the experimental arrangement of the boundary variables $X_2, \ldots, X_n$ is such that it does not affect the X_1–Y relationship in a relevant manner.

However, notice that not everything that is imperfectly understood needs to be under test. For instance, one can test the efficacy of a drug without a detailed understanding of the epidemiological characteristics of a disease. The efficacy of the drug, rather than the process of infection, is the target of such research. The point is that we must either know that the effect of the drug does not bear on the process of infection or make sure that the infection is replicated in the experiment in a way that mirrors the real-world process as closely as possible.[11] Similarly, you can do experiments on market behavior even without a thorough understanding of the mechanisms of individual choice and belief formation. Market institutions, instead of individual behavior, are under test in these experiments. Subjects may trade at a certain equilibrium price because they are acting in a fully rational way, or perhaps because they are following some rule of thumb, or even by sheer imitation. Whatever the real causal process, we can use laboratory tests to study selected aspects of specific real-world economies as long as we are confident that the same (unknown) basic principles of behavior apply in both cases.

Although both experiments and simulations are knowledge-producing devices, the knowledge needed to run a good simulation is not quite the same as the one needed to run a good experiment. A simulation relies on the assumption that the structure of the target is known, and known to be analogous to that of the simulating device. As in an experiment, then, one can learn something new by simulating the effect of certain changes in the initial conditions, which for some reason, cannot be performed

---

[11] In Chapter 9, I mentioned a famous error inferred from artificially reproducing the infection of the poliovirus in laboratory monkeys. Experiments on monkeys induced researchers to believe that the virus attacks the nervous system via the olfactory nerves, and therefore the disease should be fought by means of nasal sprays. As a matter of fact, humans are infected via the mouth, and the sprays proved to be useless (cf. Paul 1971, Chs. 12 and 23).

analytically. The obstacle in an experiment, in contrast, is not analytical in character: what is unknown is precisely those structural characteristics of a system that are taken for granted in a simulation. The advantage of experimental research is that one does not have to specify in advance the full causal principles governing the target system. The trick is to *make sure* that the target and the experimental system are similar in most relevant respects so as to be able to generalize the observed results from the laboratory to the outside world. Experimenters make sure that this is the case by using materials that resemble as closely as possible those of which the parts of the target are made. They also make sure that the components of the experimental system are put together just like those of the target, and that nothing else is interfering. Of course, quite a lot of knowledge is required in order to do so, but no fundamental theory of how the target system works is needed. Parts of the laboratory system can be put between brackets and used as "black boxes." The same processes, the same causal principles are supposed to be at work in both cases. Experimental systems are reliable if they are made of the same stuff as real-world economies. No process of abstraction from the material forces at work is needed in order to draw the analogy from the laboratory to the outside world. One may abstract from "negligible" causal factors but not from the basic processes at work. The similarity is not merely *formal* but holds at the *material* level as well.

Models, experiments, and hypothesis testing

The mediators view has some nontrivial implications. One of them concerns the way in which theory testing has to be conceived. Another one regards the sort of work that has to be done in order to make the experimental results applicable to real-world problems. In this section, I tackle the first issue and leave the second one for the last part of the chapter.

There is a very common misconception (common among philosophers and scientists alike) concerning theory testing that is most easily illustrated by means of an example. Consider a standard one-shot prisoner's dilemma (PD) game like the one in Table 10.1. Many experiments have been performed on the PD game (both one-shot and repeated), and the results are normally interpreted as providing a "refutation" of the standard theory (cf. e.g., Dawes and Thaler 1988; I discussed the results briefly in Chapter 2). This reading is behind much contemporary research aimed at finding alternatives to the standard model of rational choice or at exploring the factors that may be responsible for the anomalous behavior.

Table 10.1. *A Prisoner's Dilemma Game*

	Left	Right
Up	5, 5	0, 10
Down	10, 0	2, 2

Yet, I have put "refutation" between quotation marks for two reasons: one is that it is not clear whether one can really refute a model like the PD game. The other reason is that it can be argued that the experiments testing the PD game are not tests of this model at all.

Let us discuss the second argument first. It is often claimed that an experiment genuinely testing the PD game must make sure that the initial conditions stated in the PD game are adequately instantiated in the laboratory setting. The numbers in the standard PD matrix are to be interpreted as *utilities*, that is, as representations of the preference structures of the agents in the game. When the game is played in the laboratory, experimenters usually translate the utilities straight into monetary payoffs – for example: 2 "utils" = 2 (or 20, or 200) euros, 5 utils = 5 euros, and so on. This is based on the hypotheses (ubiquitous in applied economics) that subjects (1) prefer more money to less and (2) do not care about anything else but their own monetary gains. Then, the argument goes, if the experimental subjects do not seem to play the dominant strategy (Down, Right) this is because assumptions (1) and (2) are false in this particular experimental context, not because the PD model is false. Subjects *are* maximizing their own utilities, except that their utilities are not as experimenters assumed them to be. They are not playing the PD game, and therefore the experiment certainly cannot provide a refutation of it.

The usual way to rebut such arguments is to ask, *What* would constitute a test of the PD model then? The PD, like any good model, is a consistent description of an abstract situation. By logical analysis, we already know that if *all* the assumptions of the model are true, then the consequence must also be true. If we make sure that all the assumptions of the model are instantiated in the laboratory (that subjects are rational utility maximizers, that they have perfect knowledge of the situation, that they do not make mistakes, etc.), we are running a theorem not an experiment. The point of an experiment is to learn something we did not already know.

One could reply that there is still something to be learned by running experiments in which the postulated preferences have been instantiated: one can test the hypothesis that the subjects are *rational* or that their actions follow from their preferences and beliefs. This interpretation of game theory experiments has been defended recently by Jorgen Weibull (2002) and seems to make sense of some experiments aimed at testing complicated models, in which the rationality assumption is a tricky hypothesis indeed.[12] However, it does not seem to capture experimenters' motivations when they test relatively simple models like the PD game. Once the preferences are defined, the game is utterly trivial (especially in its one-shot version), and testing the rationality assumption does not seem a particularly interesting endeavor. A more charitable reading is that the main goal of such experiments is to test the assumptions made on the contents of individual preferences: that human beings are selfish maximizers of their own utility and that utility varies with money only. To claim that the goal is to test the rationality assumption seems to beg the real motivations behind this research program.

In the past, Ken Binmore has taken a more radical stance than Weibull:

[those who are anxious to deny that people seek only their own narrowly conceived selfish ends] argue that the players may care about the welfare of their opponents, or that they may actively want to keep their promises out of feelings of group solidarity or because they would otherwise suffer the pangs of a bad conscience. Such players will *not* be playing the Prisoners' Dilemma. They will be playing some other game with different payoffs. [...] The critic may respond that the game theorist's victory in the debate is at best Pyrrhic, since it is bought at the cost of reducing the propositions of game theory to the status of "mere" tautologies. But such an accusation disturbs the game theorist not in the least. There is nothing a game theorist would like better than for his propositions to be entitled to the status of tautologies, just like proper mathematical theorems. (Binmore 1992, pp. 313–4)

One way of interpreting this claim is that you cannot "test a model" any more than you can test *Alice in Wonderland*. In fact, it is not surprising that people find it difficult to state exactly what a test of the PD game is supposed to look like. What you can do, though, is to test an *application* of a model, a hypothesis stating that certain elements of a model are approximately accurate or good enough representations of what goes on in a given empirical situation. This point is important and worth spelling out in more detail.

Remember that in the semantic view, models come with some hypothesis attached, stating their applicability to certain real-world systems. The models by themselves do not tell you where they can or should be applied:

[12] I should thank both Dan Hausman and Chris Starmer, who helped me to clarify this interpretation of game theory tests (see Starmer, in press, and Hausman, in press).

although economic models are often made of "firms," "consumers," "markets," and so on, these are not the firms and consumers and markets of our everyday world and thus are not necessarily applicable to it.[13] According to the semantic view, the *hypothesis* says where, how, and to what extent the models may be used to understand or predict empirical phenomena. However, notice that the empirical hypotheses are much more ephemeral entities than models. The models are fairly stable and identifiable – they are the things we find in textbooks and scientific journals. The hypotheses, in contrast, are often only vaguely specified and tend to change in time as more information is gathered about the applicability of certain models to certain domains. Scientists are pragmatic people, and although some paradigmatic applications are considered more important than others, a model is always useful to a degree, as long as it is applicable to *some* situation (or, more precisely, as long as it is more helpful in understanding a certain situation than are other rival models). The fact that a model turns out not to work under certain circumstances does not count as a refutation of the model but only as a (failed) test of its applicability in a given domain.

Let us go back to the PD game now. The model is usually presented as providing a possible explanation for a vast array of real-world situations, from oligopolistic collusion to blood donation, pollution, the formation of social contracts, and so on. However, in each of these applications, the game has to be interpreted, and part of the interpretative problem consists of stating empirical assumptions like the ones we have seen above: that the players are selfish, that they maximize monetary gains (or their health, or whatever), and so on. These assumptions allow the testing of the hypothesis that the model is relevant to understanding *that* particular situation. Experimental PD games as they are normally designed seem to capture some of the features of those many economic situations for which the PD game is prima facie an explanatory candidate. That's one reason why the laboratory games arouse some interest in the first place – because they seem to be the sort of situation in which the PD game could (or should) work.

What do we learn, then, when the plot in the laboratory does not unravel as in the model? We learn, first and foremost, that the model cannot be used *that way*, that the hypothesis that players are rational maximizers, that they care only about their own money, that they have perfect knowledge of the situation, and so on cannot *all* simultaneously

---

[13] On this point, see also Cubitt (in press).

be true *under those particular circumstances*. The model is not "falsified" by the experiment (what would it mean to "falsify" an abstract entity in the first place?), its application to a particular case is. Of course, two questions then arise naturally: (1) If it does not work in this situation, does it work under other real-world circumstances that seem prima facie to be very similar to it? and (2) Why is it that the model cannot be applied this way, in this particular setting?

Question (1) should be very familiar by now. As I have argued, it cannot be solved except by checking empirically what substantial differences, if any, there are between the target systems in the real world and the experimental setting. Notice that in doing this, in a way we are treating the experiment as a rival to the standard model: we are asking which one is more similar to other situations that we are interested in understanding. We are asking whether the experiment is a better "model" for these situations than the standard model itself.

Question (2) is partly related to this issue, and partly independent. It is independent in the sense that the focus here is the laboratory system itself, rather than the external validity of the result. However, it is also related because by understanding exactly what is going on in the experiment, we facilitate the application and extension of the result to other conditions.[14] The problem here is that unlike a simple and relatively well-understood theoretical model like the PD game, a laboratory system is not a completely transparent entity. Whereas in the model we know what follows from which assumptions, in the PD experiment one has to figure out why, for example, a substantial portion of subjects plays the cooperative strategy. For this reason, a lot of experimentation is devoted to testing very specific hypotheses about particular aspects of the design: whether anonymity, the level of payoffs, the assignment of social roles, repetition, and many other factors significantly influence the phenomenon produced in the laboratory.

Thus, to sum up, the idea that certain key assumptions *must* be instantiated in the laboratory in order for it to count as a "proper" test of a model is misguided. A model is an entity made of many components, each of which may or may not be a good counterpart to what goes on in a real situation. In addition, other extratheoretical assumptions must

---

[14] The need to answer question (2) also marks the difference between a realist and an instrumentalist attitude. The realist is not satisfied by merely saying that a model applies in situation X but not in Y. This asymmetry raises a cognitive dissonance and has to be explained properly. See also the discussion in Chapter 7 of the methodological role of universality as a requirement of scientific laws.

be made in each case in order to make the model applicable to a specific real-world system. What you test, then, is a specific application of a model to a specific situation or class of situations. A negative result should not provoke panic or a retreat toward a purely formalistic interpretation of economic theory. We learn little by little, and negative results provide important clues about the limitations of the tools we devise – theoretically and experimentally.

New (and robust) phenomena[15]

Let us now turn to the second major implication of the mediators view. If economic research does not end in the lab, it follows quite naturally that economists should invest more time and effort in showing that their experimental results can be generalized to real-world contexts. It is not enough to make sure that the initial conditions of some theoretical model are instantiated in the experiment. External validity problems should be solved by combining laboratory and field evidence in the appropriate way. Economists, however, rarely practice this sort of empirical research, and the cases I have discussed in this book are representative of just a handful of attempts in this direction. Experimental economists tend, for the most part, to behave as if they were "pure" laboratory scientists.

However, we should be careful not to condemn this attitude too hastily. If experiments are in many respects similar to models and models are of different types – theoretical and applied, for example – it seems reasonable to expect that experiments also vary according to their degree of proximity to the real world. Thus, on the one hand, economists should be encouraged to invest in the underdeveloped art of applying experimental results, because that is what scientists are *ultimately* expected to deliver. External validity inferences require the specification of a target, the collection of data about the target, and the skillful combination of experimental and field data. On the other hand, an experimental result that has not been exported to the real world (yet) is not necessarily a useless experiment. The experimenter has the possibility of learning something of wider applicability than a purely laboratory game, even though she is presently unable to specify exactly the domain of application of her result. But what exactly can she learn?

[15] Some of the ideas presented in the last part of this chapter were developed jointly with Luigi Mittone (see Guala and Mittone 2002).

A most valuable feature of laboratory experimentation, one that makes it almost unique in the field of social science, is that it sometimes leads to *the discovery of new, unexpected phenomena*. Moreover, unlike field observation, laboratory work usually allows the demonstration that (1) the phenomenon in question is real and not just a spurious regularity or an artifact of statistical analysis and that (2) it is *robust* to changes in background factors. A number of phenomena discovered in the lab have passed tests (1) and (2) and are now firmly established in the economics literature: violations of individual rationality like the Allais paradox or preference reversals, but also aggregate effects like the convergence of double oral auctions toward equilibrium, the decay of contribution in public goods experiments, and so on.[16]

Sometimes phenomena are anticipated by a formal theory or a thought experiment – for example, when we try to imagine the reaction of experimental subjects placed in certain counterfactual conditions. However, phenomena are also frequently discovered by chance or noticed post hoc while analyzing data collected for different purposes (the preference reversals phenomenon is a case in point). This fact, somehow paradoxically, improves rather than affects for the worse its robustness credentials. The reasoning goes as follows: an experiment is usually aimed at testing the effect of a series of factors or independent variables $(X_1, X_2, \ldots, X_n)$ on a dependent variable (Y). Usually, the experimenter tries to design an experiment such that no other factor besides $X_1, \ldots, X_n$ is likely to have an influence on Y. (This is why abstract designs facilitate experimentation.) Then, one factor (say, X_1) is varied while the others are kept constant, and the procedure is iterated for the other $X_2, \ldots, X_n$. The list of potentially relevant X_i may come from theory, from previous experimental results, from practical insight, or just from common sense. In most laboratory experiments, we *know* that many variables have been constructed "artificially," and we are aware of the limitations of the design with respect to the real thing. The unexpected effect, in contrast, may strike us as a really genuine occurrence. The idea is that if $X_1, \ldots, X_n$ are *really* the only variables that were artificially constructed by the experimenter, then the unexpected residual effect is likely to be the consequence of some other, non–purely experimental factor.

A physical analogy may help here: cosmic microwave radiation was first observed in 1964 by Arno Penzias and Robert Wilson, two scientists at

[16] Sugden (in press) calls them experimental "exhibits"; I use here the standard terminology (*effects, phenomena*) for simplicity. See also the discussion of Sugden's exhibits in Chapter 3.

Bell Labs, while working on a problem of telecommunication technology. The isotropic radio background is a leftover from the big bang and fills the space everywhere in the universe (except where it has been "shielded" or neutralized). Regardless of where you are, it is there, although its properties may in some circumstances be difficult to detect because of disturbing factors and other local circumstances. Phenomena of this sort often emerge as residuals that cannot be imputed to the experimental procedures or other known factors, and prove to be extremely robust to measurement and experimental manipulation.

The analogy with unexpected experimental phenomena in economics goes as follows: first you observe something that you don't think has been created by the experimental procedures; then, by checking the robustness of the phenomenon to changes in background conditions, you become more confident that the phenomenon is indeed a general feature of human decision making. The checking is important because the whole inference rests on a crucial assumption: that no other "artificial" factor besides $X_1, \ldots, X_n$ has been inadvertently built into the experiment. This assumption is credible if the experiment has been designed with enough care and depends in part on the experience of the experimenter and her detailed knowledge of her system. However, no matter how experienced the experimenter is, some checking is necessary, and the scientific community will not be convinced until the attempts to "make the effect go away" have failed (Galison 1987; see also Chapter 6).

Notice, however, that the generalizability of a robust result to *specific* instances remains an empirical conjecture, which has to be further validated via case-by-case empirical investigation. By establishing robustness, the experimental economist merely points to the existence of a phenomenon that is likely to be relevant to the policy maker. Although she cannot guarantee that the phenomenon actually will be relevant in a specific case, because the effect may be neutralized by some context-specific factor, she signals a possibility. The actual applicability in each case will depend on a number of features of the specific economic system at stake (the target system).

Robustness versus external validity

Consider again the analogy with scientific modeling. A good applied model must be constructed with a clear and realistic picture of the target economy in mind. Before constructing such a model, it is usually necessary to collect quite a lot of information about the relevant characteristics of the economy one is studying (e.g., think of applied models in industrial organization). A lot of modeling, however, does not proceed that way. Instead of focusing on one specific real-world economy, the modeler may examine a set of factors or features that are likely to be relevant *generically* to a nonempty but not necessarily well-specified set of economies. One can question whether the term *applied* modeling is appropriate for such exercises. True, by adding details to the most basic theoretical models, one in a way proceeds toward a level of analysis that is more "concrete" and closer to application. But such "middle-range" models can rarely be applied directly to the functioning of a specific economic system (i.e., unless they are further modified to take into account more context-specific factors).

One way of characterizing such modeling practice is that it provides a test of the *robustness* of a result to changes in some properties of a model or set of models. In the case of experiments, there is an analogous activity of robustness testing, which falls somehow between the most abstract experiments reproducing the assumptions of theoretical models and the applied experiments used to draw external validity inferences. Indeed, robustness testing should be kept well separated from external validity. The main difference has to do with the absence/presence of a concrete, specific target system: whereas external validity requires the identification of such a target, robustness arguments do not.

When this difference is not adequately appreciated, experimental debates tend to generate some confusion. A typical case is the debate on preference reversals, a phenomenon I have already discussed extensively in Chapters 5 and 6. The research on preference reversals (PRs), has, broadly speaking, gone through two stages. A few years after Lichtenstein and Slovic's initial findings, economists started to devise experiments in order to test the reality of reversals *within the laboratory*. They tried, in other words, to check whether the anomalous evidence (the observed price-choice reversals) was merely an artifact of the experimental techniques used to elicit agents' preferences. We have also seen that economists nowadays generally agree that PRs are a real laboratory phenomenon rather than a mere illusion of the instruments of observation. The second phase of research began when experimenters turned their attention to the robustness of reversals *outside* the laboratory. They started to investigate whether PRs should be classified as "artifacts" in a different sense: whether they have been created by the experimental procedure (just like, say, superconductivity is a real physical but scientist-made phenomenon; see also the distinction between different kinds of artifacts outlined in Chapter 5).

The issue had been raised right from the start of the PR debate, to be sure. In their influential paper, Grether and Plott argued that

> The key question is, of course, whether [PRs] should be of interest to economists. Specifically it seems necessary to answer the following:
>
> (1) Does the phenomenon exist in situations where economic theory is generally applied?
> (2) Can the phenomenon be explained by applying standard economic theory or some immediate variant thereof? (Grether and Plott 1979, p. 624)

The first point inspired the work of Berg, Dickhaut, and O'Brien (1985), who tried to test the robustness of the PR phenomenon to allegedly more "realistic" conditions. The fundamental idea guiding their work is a familiar one: economic relations should not be taken as entirely general and exceptionless. On the contrary, they are relationships that hold only where the "right" conditions are instantiated. Thus, if we have a vague idea of what such circumstances may be, we can try to create an environment in which they should hold and see whether these background conditions make any difference to the anomalous phenomenon in question.

> It remains an open question whether any mechanism, particularly one which would exist in situations where economic theory is generally applied, can substantially reduce or alter [PR] inconsistencies. The mechanism considered in our work is an arbitrage procedure. In general, the possibility of arbitrage in a market setting leads to the conclusion that there cannot be market inconsistencies such as two prices for the same commodity. (Berg et al. 1985, p. 33)

The background mechanism responsible for the principles of consumer theory to hold is illustrated as follows:

> If preference reversals exist in an exchange setting, they create an arbitrage opportunity. A subject having been arbitraged is expected to realize that inconsistencies will be exploited and therefore to reduce both the rate and size of reversals. (ibid., p. 34)

Notice that the background mechanism is not formally modeled or incorporated into the theory of consumer's behavior. Economic relationships are supposed to hold across a certain range of situations, which vaguely define the domain of their application. The reasons why they hold, however, are rarely fully specified. Economists rely on various informal accounts (typically, evolutionary or arbitrage stories) of why one should be confident that some relationship holds in a certain domain. To define these stories as "theoretical arguments" is somehow exaggerated, because they seldom take the form of rigorous theories, let alone an axiomatic form.[17] Still, they provide arguments for the robustness of a relationship inside a certain domain by pointing at some background circumstances that would allow the relationship to hold. And they provide hypotheses that can be tested experimentally – no formal theory is required for that. Experimentalists are just interested in checking whether when the arbitrage mechanism is at work, reversals tend to disappear. The standard models could then be applied the way they are, provided we keep in mind their limited domain of application. Experimenters are not looking for a theoretical explanation of the mechanism, which is therefore confined in the "background" of the theory.

The design of the Berg-Dickhaut-O'Brien experiment is far from trivial. One problem with these experiments is how to combine the exchange task needed to money-pump inconsistent subjects with the already rather complex machinery of a standard PR experiment. The BDM procedure, for example, can be used to determine the *real* reservation price in the exchange mechanism only if subjects are assumed to be constantly and absolutely risk averse; otherwise, their buying and selling prices will be different (cf. Berg et al. 1985, pp. 34–35). Another incentive procedure specially invented by O'Brien to solve this problem was therefore used in the experiment. After the announced prices were elicited, the subjects were required to trade with the experimenter on the basis of their announced prices and choices.

The design also controlled for another variable, that is, repetition of the experiment. The central idea is that a period of learning may be necessary to acquire the decision skills posited by the standard models of rational choice. By controlling for arbitrage and repetition, Berg and his colleagues discovered that PRs do not disappear under these conditions. The frequency of reversals was even slightly higher when subjects experienced arbitrage than when they did not. Their dollar magnitude was, however, substantially decreased (from a mean value of 4.10 to 2.52 dollars). Repetition definitely diminished both the frequency and the dollar magnitude of reversals – the number of reversals per subject dropping from 36 percent in the first trial to 27 percent in the second, and the value from a mean dollar magnitude of 4.02 to 2.58. Not

---

[17] Some economists argue that more effort should be put into the attempt to formalize the evolutionary stories. See e.g., Nelson and Winter (1982).

surprisingly, the two effects combine so that the most significant reductions can be observed in groups subject to *both* repetition and arbitrage. Berg and his colleagues eventually recognized that the phenomenon "did not go away" but was significantly eroded.

The results of this experiment encouraged further research. Chu and Chu (1990) a few years later devised a variant of this experiment that controlled for the effects of *complexity* and of *repeated arbitrage*. They simplified the standard PR design and exposed their subjects to a series of money-pumps. Whereas simplification alone did not seem to have great impact on the phenomenon, repeated arbitrage did (most subjects required only two or three rounds of money-pumping in order to revise their preferences). Moreover, the effects of learning were relatively persistent: once exposed to repeated money-pumping, subjects acted more consistently with standard economic theory in immediately subsequent tasks.[18]

These experiments, then, showed that preference reversals could be reduced in certain situations. But are these preventatives active in real-world markets? It is fair to say that the economists engaged in PR research did not go very far in investigating external validity. Chu and Chu summarize their results by saying that the PR phenomenon appears to be vulnerable to a "marketlike" environment (1990, p. 910), but such a claim is at least ambiguous. The argument seems to go as follows: "real markets" involve repeated choice, under specific institutional rules, and are populated by arbitrageurs. Therefore, by making the classic PR experiment more similar to "real markets," we test the external validity of the PR phenomenon. I have put "real markets" between scare quotes because not all *real* markets are of this kind. The real estate market, for example, is populated by many traders who will not engage in that sort of transaction more than once or twice in their whole life. The price is often determined by a first-price sealed bid mechanism, and most traders do not have a chance to learn that their preferences are inconsistent by being repeatedly money-pumped (fortunately, one might say!). Thus, the external validity of the experiments used to test the robustness of PRs will not stretch to these circumstances. Their results will be applicable only

---

[18] Knez and Smith (1987) and Cox and Grether (1996) describe similar attempts to test the robustness of reversals in marketlike environments. Because the interpretation of these experiments is rather complex and from some respects controversial, I shall not comment upon them here.

to real-world economies with repetition, arbitrage, and the English auction mechanism. The "target" in this case is defined implicitly (and only roughly) by the features of the experiments.

This is quite typical of robustness tests: instead of mimicking a specific real-world system, such experiments define the domain of their results implicitly and somehow generically. But external validity inferences do not have much bite unless one systematically investigates the degree of similarity and dissimilarity between laboratory and target systems. In the PR experiments, the experimenters had an abstract rather than concrete target in mind: an "ideal" competitive market with repetition and arbitrage. Of course, their results are not useless – they are indeed extremely useful to test the *robustness* of the PR phenomenon. (They do so in a negative way – by indicating where the phenomenon may break down, rather than by showing us that it occurs in a certain set of real-world situations.) The distinction between robustness and external validity is crucial to highlight the confusion here.

The library of phenomena

Unlike some of its neighbor disciplines, such as experimental psychology, experimental economics grew within (and had to defend itself from) a scientific paradigm that attributes enormous importance to theory. This is probably why it was sometimes easier and more effective from a rhetorical viewpoint to present experimental economics as primarily devoted to theory testing. This view is mistaken, because the proper role of experimental economics is to mediate between abstract theory and concrete problem solving in the real world. In many respects, experiments resemble models, for they are systems that are artificially isolated from the noise of the real world – but with the added bonus of a higher degree of concreteness.

Like models, experimental results must eventually teach us something about the real world. However, many experimental results in economics are never applied to real-world situations. In some happy cases, the experimenter can go all the way from the model on the far left to the target system on the right of Figure 10.2, but these cases are quite rare. Most cases of experimentation involve inferences to generic circumstances rather than to specific targets. This may be either because the target is willingly left unspecified or because it cannot be studied properly for lack of data. Experimental economists nevertheless help the applied scientist by compiling a *library of phenomena*: a list of mechanisms, effects, and biases that may be relevant in concrete applications. In order to apply this knowledge to some real-world situation, it is necessary to examine the specific characteristics of the target domain and based on this local information, evaluate the relevance of the phenomena found in the "library" case by case.

This way of framing the problem has the advantage of recovering a basic distinction, between pure and applied science, while defending experimental economists from the charge of pursuing futile research. Although a research program should eventually produce applicable results (that's what scientists are paid for, after all), it needs not do so for each single experiment. A single experiment may just highlight a phenomenon or mechanism, to be later exploited by applied scientists when they deal with specific cases. To export a phenomenon in the real world requires detailed knowledge of the domain of application. Because the required knowledge is context specific and probably generalizable only up to a point, it is reasonable to have a division of labor between applied scientist and experimenter (which of course does not mean that in some cases, the same person cannot play both roles at the same time).

ELEVEN

On Monetary Incentives

Some experimenters may find it surprising to see the issue of incentives relegated to the last chapter of a methodology book. Monetary incentives are at the center of most methodological controversies in experimental economics and have sparked some heated exchanges among practitioners, so one would perhaps expect them to have a more central place in a book like this.[1] The main reason to delay the discussion until now is not that I find the incentives problem uninteresting or unimportant, but that it is a complicated one. It can be tackled properly only once the right conceptual tools are available; given that I provided the tools in the previous chapters, I can now put them to work.

The idea of using monetary rewards sometimes generates hilarity among noneconomists ("These guys pay their subjects to behave like economists would like them to behave!"), whereas the absence of incentives is dismissed by economists equally bluntly ("What can you learn from 'cheap talk'? Put your money where your mouth is!"). But more importantly, the presence of "adequate" monetary incentives (we shall see what *adequate* means shortly) has become de facto a prerequisite for publication in economics journals – and, conversely, the lack of incentives is considered a sufficient condition for the rejection of an experimental study. In contrast, social, cognitive, and economic psychologists tend to apply a less rigid policy. Many experiments in these areas are performed with incentive structures that would be considered inadequate in economics and often lack monetary incentives altogether.

---

[1] Cf. e.g., Harrison (1989) and the subsequent debate in the *American Economic Review* (1992). Cf. also Harrison (1994) and Hertwig and Ortmann (2001), which will be discussed in more detail later.

As with many other issues discussed in this book, there is probably an interesting sociological story to be told here. The pioneers of experimental economics faced the problem of distinguishing their work from the research carried out in neighbor disciplines. At the same time, it was also useful to differentiate experiments from other economic methods of observation, such as surveys and contingent evaluations. Incentives (together with, e.g., the semitheoretical framework adopted in Vernon Smith's early methodological papers) probably served this differentiating function, at least in part. The legacy of this function is still evident from the fact that the issue of incentives is often couched in terms of "the economics-psychology methodological divide." I do not intend to review this more general debate here;[2] however, following the predominantly normative approach of the book, I will try to clarify the narrower and yet highly complex issue of incentives – an issue that, once analyzed in depth, turns out to involve a number of methodological problems that are often inappropriately conflated. The main conclusion will be that once the motivations and the arguments that underlie monetary incentive norms have been unpacked, one can appreciate how the question, Do incentives matter? can be answered only in a context-specific manner. Do they matter ... *for what*?

The 'precepts'

Early economic experiments (even "paradigmatic" ones, like Smith 1962 or Allais 1953) lacked what contemporary experimental economists consider an "adequate incentives structure." The norms regulating financial incentives were codified later, in a series of papers written in the late seventies and early eighties by Vernon Smith (1976, 1982) and Luis Wilde (1980).[3] Incentives are discussed in four of the five so-called *precepts* of experimental economics:

1. *Nonsatiation*: choose a medium of reward such that of two otherwise equivalent alternatives, subjects will always choose the one yielding more of the reward medium.
2. *Saliency*: the reward must be increasing in the good and decreasing in the bad outcomes of the experiment.

---

[2] But see Cox and Isaac (1986), Hogarth and Reder (eds. 1986), Smith (1991b), Loewenstein (1999), and Rabin (1998, 2002).

[3] The idea of using monetary rewards was borrowed, somehow ironically, from the work of two psychologists (Fouraker and Siegel 1963).

3. *Dominance*: the rewards dominate any subjective costs associated with participation in the experiment.
4. *Privacy*: each subject in an experiment receives information only about her own payoffs.

The fifth precept (*parallelism* or external validity, as I call it throughout this book) is also partly associated with the issue of incentives, but I shall bracket it for now and come back to it later on. The precepts form the core of "Induced Value Theory" (Smith 1976) and are to be interpreted as "a *proposed* set of *sufficient* conditions for a valid controlled microeconomic experiment" (Smith 1982, p. 930, my emphasis). Notice two things: first, the conditions identified by the precepts were not intended to be necessary ones; that is, according to the original formulation, a perfectly valid experiment may in principle be built that nevertheless violates some or all of the precepts. Second, the precepts should be read as hypothetical conditionals ("if you want to achieve control, you should do this and that"), and should emphatically not be taken as axioms to be taken for granted. "The truth of these precepts can only be established empirically" (Smith 1982, p. 930, n. 10). Consider also that the precepts provide broad guidelines concerning the control of individual preferences, which may be implemented in various ways and may require ad hoc adjustments depending on the context and the particular experimental design one is using. In fact, money or financial incentives are never mentioned in the precepts. The principles state only in abstract terms what kind of properties an appropriate reward medium should have but do not say what the medium should be. Money may be one way of implementing the precepts, but is not necessarily the only one. In light of the fairly rigid interpretation that has become prevalent in experimental economics, the Smith-Wilde precepts appear distinctively liberal in their original formulation.

So what are the precepts for? A major problem with field research is that some key variables of economic theory, such as agents' preferences, are not directly observable. If you are interested in explaining, say, price variations in a market, you have to derive the demand and supply schedules (two crucial explanatory factors) from other observable variables, based on auxiliary assumptions that are usually as difficult to test as the main research hypothesis (that markets equilibrate at efficient prices, say). Subjects' preferences and beliefs are directly unobservable in laboratory experiments too, of course, but can be more easily detected and controlled therein. The precepts are a set of guidelines intended to achieve control of this sort. They were originally supposed to apply to one particular type of economic experiment: the market experiments at the core of Vernon Smith's early research.[4] In fact, in his 1976 paper on Induced Value Theory, Smith states explicitly that the principles apply "to experiments designed to test price theory propositions conditional on known valuations. Separate experiments can be designed to test propositions in preference theory" (Smith 1976, p. 275).

Most of Smith's early experiments were aimed at investigating a crucial aspect of market phenomena that was left remarkably obscure by existing theory. The experiments focused, in particular, on the role played by the institutions that coordinate individual behavior in competitive markets. In Smith's own terminology – borrowed largely from the mechanism design theory of the sixties and seventies – a "microeconomic system" is analyzable into three major components: the environment, the institution, and the outcome (the behavior of the agents in the market). The outcome is modeled as a function of the environment and the institution. The institution is basically (I'm simplifying here) a set of rules governing behavior by setting incentives, punishments, and their enforcement. The environment is a complex set of factors including the commodities to be exchanged, the agents in the market, their individual endowments, their utility functions, and the technology (costs).

In order to learn about the effects of these factors on the outcome behavior (e.g., the sort of prices that are generated in a market defined by a certain environment and a certain kind of institution), the ability to control preferences is quite crucial. By controlling preferences, first, one can try to systematically vary the supply/demand schedules in a given institution and observe the results of such variations. Second, one can keep the preferences fixed ("in the background," using the terminology introduced in the first part of the book) and observe the effect of using different institutions in a given environment (cf. Smith 1982, p. 927). The precepts provide some guidance on how to achieve effective control of subjects' preferences. A typical application works as follows. Suppose you want to induce in your experiment simple supply and demand schedules like those in Figure 11.1. The customary way of achieving this goal is to assign your subjects some definite roles in the experiment, by dividing them in two groups of buyers and sellers with well-defined reservation prices. The reservation price of a seller can be interpreted as the cost of production for each unit of the exchange good. The reservation price of a

---

[4] Cf. e.g., Smith's seminal (1962) article as well as the papers collected in Smith (1991a).

Table 11.1. *Simple Incentive Structure*

No. of Subjects	Reservation Price
10 sellers	30 tokens
20 sellers	10 tokens
10 buyers	35 tokens
20 buyers	15 tokens

buyer can instead be seen as the price the experimenter is willing to pay each buyer for a unit of the good once the experimental market is closed.

The supply/demand schedules in Figure 11.1 can be induced by setting reservation prices as in Table 11.1 (assuming that each buyer can exchange at most one unit of the good during the experiment). Notice that the prices are expressed in experimental tokens. The key move, according to the precepts of Induced Value Theory, is to make sure that the tokens will be exchanged (privately) at the end of the experiment with some other reward medium, at a rate that satisfies the criteria set out in the precepts themselves – hence the habit of using real money in quantities that are likely to dominate all other costs of participating in the experiment.

The precepts, as noticed by various commentators (e.g., Starmer 1999), add a set of auxiliary assumptions to the hypotheses usually tested in market experiments. The interpretation of the results of an experiment (e.g., that institution *X* is more efficient than institution *Y*) relies crucially on the background assumption that preferences have actually been controlled – a hypothesis that may in principle be questioned and tested empirically on its own. This is just the familiar Duhem-Quine problem discussed

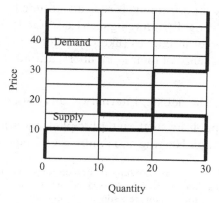

Figure 11.1. Simple supply and demand schedules.

extensively in the first part of this book. The point I would like to stress here is that these experiments are *not* the only or the typical experiments performed by economists. Following Cubitt, Starmer, and Sugden (2001), we may identify two broad categories of experiment: experiments aimed at testing the effect of individual preferences, beliefs, endowments, institutions, and so on on market outcomes (which test "$P \rightarrow Q$" type of inferences: "if agents have qualities P, then in given circumstances they will do Q") vs. experiments aimed at testing the standard assumptions imposed on individual preferences, beliefs, and so on (which test claims of the form "P": "agents have qualities P").[5] In the first kind of experiment, it is necessary to try to implement the standard assumptions of economic theory by inducing preferences, beliefs, and so on that are consistent with the assumptions of rational choice theory. Otherwise, one would not be testing the $P \rightarrow Q$ proposition at all. The aim of these experiments typically is to find out whether certain market institutions are able to aggregate individual preferences with certain characteristics in a "desirable" (e.g., efficient) way. But in order to find that out, one has to make sure that subjects' preferences have the characteristics postulated by standard microeconomic theory. In contrast, in experiments of the second type, the assumptions of individual decision theory are themselves under test: the aim is to figure out whether individual preferences (and/or beliefs) have the structure postulated by the standard models. The precepts of experimental economics lose much of their appeal in contexts like these, because clearly there is little point in trying to induce the behavior one is supposed to be testing in the first place.

Nevertheless, the norms regulating the use of monetary incentives in experimental economics are enforced in *all* cases, regardless of the kind of experiment one is performing. Is such an attitude reasonable? Are we losing something by rigidly enforcing such rules? Or do the disciplines that underplay the role of incentives (like psychology) produce less reliable results as a consequence of their more liberal attitude?

The debate: Internal validity issues

In a recent paper published in the interdisciplinary journal *Behavioral and Brain Sciences*, Ralph Hertwig and Andreas Ortmann (2001) argue that

---

[5] Strictly speaking, hypotheses on the nature and structure of preferences, beliefs, etc. should also be represented as inferences of the $P \rightarrow Q$ kind. When you are testing, e.g., the transitivity axiom of choice theory, you are testing the proposition that $[(x \succ y) \& (y \succ z)] \rightarrow (x \succ z)$. I stick to the above formulation mainly for presentational ease.

psychologists would be better off by adopting some of the methodological standards of experimental economics.[6] Their paper acted as "target article" for several commentaries by distinguished economists, psychologists, and social scientists. Because the use of monetary incentives is among the "good" norms to be exported into psychology, the *BBS* discussion offers a useful overview of the arguments for and against incentives. Moreover, Hertwig and Ortmann's proposal applies exclusively to the areas of experimental economics and psychology dealing with individual judgment and decision making, behavioral game theory, and cognitive psychology – precisely the experiments to which the Smith-Wilde precepts were originally *not* supposed to apply. It therefore provides a good test of the claim that economic incentives are always required in experimental research.

Hertwig and Ortmann (2001, p. 390) put forward four arguments in favor of monetary incentives:

1. Monetary incentives are easier to implement than other (nonmonetary) incentives.
2. Money is particularly appropriate to fulfil the nonsatiation requirement.
3. Economic theory is straightforwardly translated into experiments with monetary incentives.
4. Monetary incentives reduce the variation in subjects' performance.

It is useful to divide the arguments in two categories. On the one hand, there are purely pragmatic considerations, appealing to the fact that money rewards seem to be particularly handy for the fulfillment of the precepts. Arguments 1 and 2 fall in this category: in principle, whether we use real money, fictional money, candies, badges, or some other kind of reward might not matter. The reason real money is so widely used is that we feel confident that most people care about it, and that we all want more. (Most people, for example, tend to feel satiated after they have eaten lots of candies and don't want any more.) Money, moreover, seems to be almost universally attractive in our culture. If you are dealing with a group of computer geeks, you may be able to control their preferences by means of free copies of the magazine *PC World*, but that may not work with other people. Finally, money is something we can (and often do) consume on our own. If you use tennis rackets as a reward, you may

---

[6] Hertwig and Ortmann are advocating a process that in many ways is already taking place. It is much more common nowadays to see psychological experiments that fit "economic" requirements, sometimes because the author is genuinely convinced of their usefulness, sometimes just for the sake of appealing to a wider audience.

find out that some subjects are very keen to make sure that at least one of their friends also wins a racket (it's boring to play tennis on your own), thus displaying other-regarding preferences.

The other two arguments are more interesting because they involve more substantial philosophical presuppositions. Let me start with number three: economic theory is straightforwardly translated into experiments with monetary incentives. It is true that economic models can easily be applied to situations in which monetary payoffs are at stake. Nonexperimental economists exploit this routinely by simply assuming in their models that agents' utility functions have income as their only argument, and by adding some other ad hoc assumptions on diminishing marginal returns, risk aversion, and so on. However, they are (supposed to be) aware that these are hypotheses that may turn out to be false. The problem is that it is difficult to test them severely by means of field data. Experimental economics allows the testing of such conjectures in controlled conditions; but then why should we *impose* the requirement that money is the only relevant reward, that subjects do not care about other participants' rewards, and so on?

Recall the distinction introduced in the previous section between the two broad categories of experiments. If you are testing some hypothesis *other* than those normally imposed on preferences (say, the efficiency of some market institution), it is legitimate to try to make sure that the conditions included in (most) economic models (selfishness, consistency, etc.) are implemented. You may fail to do so (it is an empirical matter, let us keep it in mind), but you can try to do it nevertheless. In many experiments, however, we are not supposed to do that. Indeed, in the cases discussed by Hertwig and Ortmann, the goal is almost invariably that of *testing* the assumptions that economists routinely (and sometimes, alas, uncritically) introduce into their models. In such cases, is it legitimate to induce the assumptions of economic models experimentally?

The answer in a nutshell is yes, sometimes, some of them. Remember that the experimental method works by eliminating possible sources of error or, in other words, by controlling systematically the background factors that may induce us to draw a mistaken inference from the evidence to the main hypothesis under test. A good design is one that effectively controls for (many) possible sources of error (see Chapters 6 and 7). In an experiment on individual decision making, we may be interested in testing a number of different hypotheses. For example:

(a) Are experimental subjects selfish?
(b) Are experimental subjects only interested in money?
(c) Are the preferences of experimental subjects consistent?
(d) Are experimental subjects' beliefs correct?

Each of these hypotheses can be tested severely on its own, *provided that we already know the answer to the other questions.* Remember that a crucial advantage of controlled experimentation is the possibility of varying *some* experimental conditions or variables while keeping the others fixed. So if you want to test the hypothesis that preferences are consistent (c), you should make sure that you can control the objects of preference (b), the independence of individual utility functions (a), and individual beliefs (d). To illustrate, we can use a famous example due to Amartya Sen (1993, p. 501): from the mere fact that a person chooses "nothing" from the set of options $S_1 = \{$apple, nothing$\}$ but picks up an apple from the set $S_2 = \{$apple, apple, nothing$\}$, we cannot infer reliably that that person's preferences are inconsistent. Perhaps in the former case she was at a dinner party with friends, and she was being polite by not taking the last apple left on the table. It is perfectly legitimate then, *if we are trying to test a hypothesis on the structure of preferences,* to try to control or induce their *contents,* at least in part – for example, by trying to eliminate other-regarding motives or the influence of social norms by design. However, it would be foolish to try to induce *all* the principles of consumers' theory, because a genuine experiment requires that some degree of freedom be allowed in the system one is studying. This is indeed what distinguishes experiments from simulations or theoretical demonstrations (Chapter 10). If you make sure that *all* the assumptions of a theoretical model are implemented, you are not doing an experiment: you are proving a theorem.[7]

Finally, remember that not all experiments test theoretical models. Many experiments are aimed at testing the robustness of phenomena that have been established experimentally, or at exploring aspects of economic systems that cannot be adequately modeled theoretically. In all experiments that are *not* aimed at theory testing, it is obviously legitimate to explore "what happens if . . ." some extra- or nontheoretical

---

[7] Another way to put it is that experimentation involves an element of *surprise,* whereby one learns something that has not been "constructed" in the experimental system itself. On the difference between "surprising" results in experiments and simulations, see Morgan (2002).

conditions are instantiated. Again, in such cases it may make a lot of sense not to implement some of the precepts. Several experiments on social dilemmas, for instance, violate the requirement of privacy because information about other subjects' behavior and rewards is likely to be an important explanatory variable in itself. In fact, whether social dilemma games are played with face-to-face interaction, with the possibility of striking (nonenforceable) agreements, and so on does have an influence on the levels of cooperation and defection (cf. Orbell, Dawes, and van de Kragt 1990 for a survey of results).

The last argument (4) in Hertwig and Ortmann's list is that financial rewards reduce variation in subjects' performances – or, in other words, help them to find the "right" answer to the experimental task. This argument cannot be used to make a general case because several experiments are *not* aimed at testing subjects' cognitive capacities. In many situations, there simply is no right answer, so the normative theory cannot work as a benchmark against which to measure the rates of error.[8] In other cases, we are not even sure whether the theoretical prediction is *really* rational after all. Backward induction arguments in finitely repeated games are a case in point, for many economists and philosophers still find them unconvincing. Even if we see behavior converging to the predicted outcome in such games when the stakes are high, what are we supposed to conclude?

Let us restrict our attention, for the sake of the argument, to the experiments that are aimed at testing normative models of behavior. The rate of error, or the "noise" in the experimental data is in *many* such cases reduced by the introduction of incentives, as if subjects were effectively paying more attention to what they were doing and therefore following the normative theory more closely. (Cf. e.g., Smith and Walker 1993 and Camerer and Hogarth 1999 for some empirical data that support this claim.) However, notice that, first, the deviations from the normative model are usually reduced without disappearing entirely; second, the reduction does not take place in *all* cases: in the majority of experiments, incentives do not make a difference in terms of average performance (even though they reduce variation), and in a substantial minority of cases (29 percent of experiments on judgment and decision making, according to Hertwig and Ortmann 2001), they cause a *worsening* of performance.

---

[8] Cf. Hertwig and Ortmann (2001, p. 391) as well as Baron's (2001) and Betsch and Haberstroh's (2001) commentaries.

How is this possible? One of the surprising aspects of the debate on monetary incentives is that economists seem pretty happy to adopt a very important (and expensive) methodological convention without investigating the mechanisms linking incentives to experimental behavior. There is little discussion, in other words, of *why* incentives might matter (if and when they do). Following a recent paper by Daniel Read (in press), one can sketch at least three different stories about the effect of monetary incentives:

(I) The *cognitive push* story: the incentive induces the agent to think longer or harder;

(II) The *motivational rerouting* story: the incentive alters what the agent perceives as his or her goals;

(III) The *Pavlovian trigger* story: the response can only be given in the presence of incentives. (Read, in press)

Attempts to control preferences in market experiments clearly try to exploit mechanism (II): we try to make sure that the subjects care only about what we want them to care about, so that we are able to observe the effects of induced changes in the experimental environment. The third mechanism (III) sounds congenial to many economists trained in the behaviorist tradition (also known as revealed preference theory) and starts from the presumption that human beings suffer from substantial amounts of false consciousness. We all like to think of ourselves as nice, caring, altruistic beings, but then when put in the appropriate circumstances (when money is at stake), we just cannot help but act as the cynical agents postulated by economic models. If this is the case, then, asking hypothetical questions (What would you do if . . .?) or trying to incentivize by using alternative mediums cannot possibly give useful insights into real economic behavior. The first story (I), finally, is the one discussed above.

Now, one point of articulating these different mechanisms is that *they do not always necessarily "push" in the same direction*. The motivational rerouting mechanism (II), for instance, is often used by psychologists to argue that the introduction of monetary incentives may, in some cases, put off the subjects who were up to that point just trying to do well in the experiment. If the incentives really reduce or eliminate this "intrinsic motivation," the apparently inconsistent effects reported by Camerer and Hogarth and others are easily explained (cf. Read, in press). But if this is true, in turn, the case for the generalized usage of monetary incentives is *undermined* not overdetermined by the numerous arguments provided for its support. Perhaps the correct conclusion is that we need more experiments to test the impact of incentives on behavior. However, when the incentives are under test, clearly, raising the stakes does not solve *all* the problems of experimental design – indeed, in some cases, it creates new ones that need to be dealt with separately. Suppose, for instance, that we do observe more free riding in a social dilemma game with monetary incentives than in a game played with candies. What does it mean? That the players are more rational? Or that they are more selfish? Perhaps money makes people play for themselves, whereas candies signal some kind of altruistic attitude. (I'm using silly examples here on purpose, but see van Vugt's 2001 commentary for a more serious example.) Once again, a certain design is good or bad depending on what your goal is – what hypothesis you are trying to test.

The convention of not running or simply disregarding experiments without monetary rewards appears particularly puzzling when the incentive structure is the main object of study. Consider the experiments on preference reversals discussed in Chapters 6 and 7. It is not rare to hear economists saying that preference reversals were first discovered by Grether and Plott (1979) or, more sophisticatedly, that Grether and Plott performed the first "proper" (i.e., adequately controlled) experiments on reversals. This view stems precisely from the mistaken assumption that only those experiments that implement the precepts are "adequate" or provide valid results. As a matter of fact, one can learn very interesting things by intelligently violating the precepts. The obsession with incentives, for instance, led economists to disregard, among Lichtenstein and Slovic's early experiments (1971), those performed *without* the BDM mechanism. Thus, when the BDM came under attack in the mid-eighties, economists did not consider the very evidence demonstrating that the BDM was not significantly distorting the observed choices. It took many years for experimenters to notice this fact and replicate reversals with and without the BDM once again (Cox and Epstein 1989; Tversky, Slovic, and Kahneman 1990). Had they paid more attention, they would have avoided a great deal of unnecessary trouble.

External validity issues

All the arguments reviewed in the previous section revolve around issues of internal validity. External validity, however, figures prominently in the debate on incentives, in ways that are usually not clearly spelled out by the debaters themselves. In this section, I want to consider two arguments that are often presented by the supporters of monetary incentives. One is that monetary incentives help also in the observation of non–financially motivated behavior. The other one is that behavior falling in the intended domain of economic theory is financially motivated. Both arguments then conclude that monetary incentives somehow contribute to the generalizability of experimental results.

Let me start with the first one. Suppose that under some circumstances, some people's preferences turn out not to be controllable by "normal" monetary incentives. Standard one-shot social dilemma games may be one such example: many people prefer to invest money in the collective project instead of free riding, despite the fact that free riding dominates (financially) cooperation. Let us also imagine, for the sake of argument, that *any* level of monetary reward will be insufficient to induce selfish preferences in some subjects (despite the cynical dictum that "every man has a price"). The idea, endorsed by Hertwig and Ortmann and others (e.g., Roth 2001), is that if we observe cooperative behavior *despite* the presence of strong monetary incentives, then the evidence in favor of the anomalous phenomenon is just as strong or even stronger than the evidence we would have collected had we observed the same phenomenon without monetary incentives.

For example, if in prisoner's dilemma games (or public good, trust ultimatum, or dictator games) the behavior of participants does not correspond to the game-theoretic predictions, that is, if they show more altruism (trust, reciprocity, or fairness) than the theory predicts, then these findings also tell us something about the other nonmonetary motivators. (Hertwig and Ortmann 2001, p. 390)

In other words, a violation of the standard theory is strengthened by the fact that it was observed in conditions in which the theory has, so to speak, the best shot. If it fails there, how can it work in other, less favorable conditions? This argument, unsurprisingly, falls short of proving the universal validity of monetary incentives. Hertwig and Ortmann ask us to focus on the case in which the theory fails with high incentives but to consider the opposite question: what do we learn if the theory does *not* fail to predict under such "ideal" circumstances? What if, when provided with (some "appropriate" level of) incentives, most people do free ride in prisoner's dilemma-like games?

The answer should be quite obvious: by conducting experiments with incentives at level *x*, we learn that the theory works when incentives are set at level *x*. It is true that we are tempted to draw inferences to wider conditions: if people are not selfish with "high" incentives, we feel

244   *The Methodology of Experimental Economics*

somehow inclined to believe that they won't be selfish when the monetary rewards are lower (ceteris paribus). If people are selfish with "low" incentives, similarly, we feel encouraged to infer that they will also be selfish when the rewards are higher. Such inferences are based on some simple but plausible psychological assumptions, which can be tested independently. It is important to stress the testability and indeed the importance of actually testing such hypotheses, because even "plausible" hypotheses often turn out to be contradicted by the evidence. Our judgment is heavily loaded with theoretical presuppositions, and in fact, scientists working in different theoretical paradigms often have different intuitions regarding the generalizability of the same results. A well-known case concerning the effects of monetary incentives is the controversy on blood donation, which in the 1970s saw sociologists, psychologists, and economists fighting on various fronts. Several distinguished economists could not even conceive of the possibility that monetary incentives could reduce donations – a fact that is easily explained by alternative sociological theories (Fontaine 2002 provides a comprehensive and fascinating historical reconstruction). However, even in experimental economics, incentives sometimes turn out to interact in surprising ways with other factors. In preference reversal experiments, for instance, higher incentives seem to *increase* the frequency of reversals (Grether and Plott 1979). And as noticed by Read (in press), reported "reductions" in the frequency of certain anomalies of decision making often result from comparing experiments with *high hypothetical* monetary payoffs versus experiments with *low real* monetary payoffs. However, anomalies like the Allais paradox exploit distortions in the perception of risk that are stronger when the payoffs are *large*. To generalize from a reduction in the low-incentives case to a further reduction in the high-incentives case seems unwarranted here.[9]

The general point is that whether an experimental result can be generalized or not depends on at least two sets of conditions: the mechanism

---

[9] In the classic Allais paradox experiment (Allais 1953), subjects are asked to choose first between (*A*): one million for sure and the lottery (*B*): five million with probability 0.10, one million with probability 0.89, or 0 with probability 0.01; then, they are asked to choose between (*C*): five million with probability 0.10 or 0 with probability 0.90 and (*D*): one million with probability 0.11 or 0 with probability 0.89. Many people choose *A* and *C*, thus violating the independence principle of expected utility theory. The .01 chance of "losing" one million in option *D* appears psychologically very relevant in the first choice task but is considered irrelevant in the second, in which there is a big chance of winning nothing anyway. Clearly, the higher the stakes, the more relevant the .01 chance appears to be, which explains why Allais-type violations are eroded when the stakes are small (as in, e.g., Conlisk 1983).

by which the main causal factor produces the effect in the experiment and in the field, and the presence of other factors or circumstances that may interact with the main factor in the field to generate similar/different effects. Once again, the moral is that much more ought to be known about the mechanism(s) triggered by monetary incentives and about the real-world circumstances to which the experimental result is to be generalized. Notice that lacking this sort of knowledge does not make an experimental result useless. If we set the payoffs in the region of, say, a thousand euros (or dollars) and find out that people behave in a selfish way (as the theory predicts), we have a perfectly valid result (internally speaking: we know how people behave in those circumstances), which, however, cannot be automatically generalized to predict and explain people's responses when payoffs are in the region of, say, a hundred euros (dollars).

It is hard to establish that monetary incentives are good *in general*, or no matter what. In reality, a good design is one that allows the severe test of a *specific* hypothesis. In science, it is very difficult to devise an experiment that is able to test several hypotheses at once. In the first part of the book, I try to explain why this is the case, by pointing out that a good experiment requires *some* variation in the experimental conditions – but not *too much* variation, and of the right kind. If you want to test several hypotheses at once, you typically have to vary many things at once, and the concomitant variations will confound your results. Thus in order to decide whether an incentive structure is adequate for a given experiment, we need to ask, What are we trying to find out in this experiment? What is the hypothesis under test? The discussion on incentives is often confused by the fact that it is not clear what kind of experiment we are trying to do in the first place. A design is not good or bad in general. It can only be good or bad given what you are trying to achieve.

The last argument I'd like to discuss points in a similar direction. The premise of the argument is that in the intended or "proper" domain of economic theory, the financial incentives are high. Hence, it is appropriate to test economic theory under circumstances that mirror those of its intended domain of application. This argument comes up very often both in official and informal methodological discussion, but a bold formulation of it can be found in a recent article by Ken Binmore (1999). Binmore makes a more articulate claim, to be precise, and highlights "adequate incentives" as one among several criteria that an experiment should meet in order to be valid. Other criteria are that the tasks faced by the subjects be relatively simple (and transparent) and that subjects be given enough time for trial-and-error learning. I shall not discuss these other requirements (see also Plott 1995), but most of what I say here about incentives can be easily extended to them. The gist of the argument is presented by means of a chemical analogy:

> I will happily undertake to refute chemistry if you give me leave to mix my reagents in dirty test tubes. Equally, if you undertake to prove in the laboratory that young stockbrokers cannot learn their trade by denying my subjects access to the conventional wisdom that the stockbroking profession has built up over many years of interactive trial-and-error learning. Just as we need to use clean test tubes in chemistry experiments, so we need to get the laboratory conditions right when testing economic theory. (Binmore 1999, p. F17)

According to the dirty-tube analogy, experimenters must be careful to create the "right" conditions for a certain hypothesis to be tested. Suppose we are testing the hypothesis that two elements (say, aluminium and oxygen) combine in an oxidation reaction: $4Al + 3O_2 \Rightarrow 2Al_2O_3$. Now, in order for this hypothesis to be tested properly, we obviously need to make sure that the initial conditions stated on the left-hand side of the formula are instantiated in the "test tube." Otherwise, we would be testing *another* hypothesis: to observe that the experiment does not generate the compound $2Al_2O_3$ would not count as a refutation of this part of chemical theory. However, the analogy to chemistry is shaky insofar as standard economic theory does not provide a decent account of the background factors (e.g., the "conventional wisdom of the stockbroking profession") that supposedly make it applicable in a certain range of circumstances.

Notice also that in the chemistry example, the problem is one of *eliminating* or neutralizing disturbing factors ("keeping the test tube clean"). In economics, the problem is more complicated because we have to *add* to the experiment the factors that make the theory applicable. Hence, it is not even clear how the analogy should work in the social sciences. What is the equivalent of a vacuum or a clean test tube? As George Loewenstein (1999) points out, every situation is a social situation. Even the most abstract decision problem must be and is interpreted by the experimental subjects (e.g., as a game situation, an experiment situation, an exam situation, or some kind of real-life situation). "No context" is a context in itself, in social science experimentation. The real issue then is, Does the experimental context afford generalizing to other nonexperimental circumstances? This is the old problem of external validity, and as I have argued in previous chapters, it can be solved only by appropriately combining laboratory and field evidence – which raises a related problem. Suppose we were able to build into the experiment the factors that make the theory work. Even in such a case, the generalizability problem would still hold, for the experiment would not teach us anything about what happens when those circumstances do not hold. We are not all stockbrokers, and by restricting the domain of application of the standard theory to the circumstances under which agents have the chance to acquire stockbrokers' conventional wisdom is to impose a big restriction on the domain of economic theory.

By radically redefining the "intended" domain of economic theory, Binmore tries to kill two birds with one stone. Common sense suggests that internal and external validity are related by an inverse relationship (see Chapter 7, as well as Loewenstein 1999, and Harrison and List, in press). Given that the world is generally complicated and messy, an experiment that is strong with respect to internal validity is likely to be weak from an external validity viewpoint. By identifying the "proper" domain of economic theory with behavior motivated by high monetary stakes, one can pretend that incentives can contribute to achieve a higher degree of internal *and* external validity at once. However, it is obviously an illusion. One cannot make the results more generalizable by merely increasing the incentive structure. One could claim that only those real-world situations mirroring the incentive structure of the experiment are worth studying in economics departments. Here, I'm afraid, most economists will part company with Binmore's approach. Binmore is ready to bite the bullet, in contrast:

I know that denying the predictive power of economics in the laboratory except under such conditions implies that we must also deny the predictive power of economics *in the field* when such conditions are not satisfied. But have we not got ourselves into enough trouble already by claiming vastly more than we can deliver? I am certainly tired at having fun poked at me by marketing experts for supposedly believing that economic consumer theory is relevant to the behavior of customers buying low-cost items under supermarket conditions. How could customers find the time to research the value of the products on sale? Even if they could, the supermarkets would simply speed up the rate at which they differentiate their products and packaging. (Binmore 1999, p. F17)

There is some bold honesty in these paragraphs, especially in the acknowledgment that the domain of applicability of standard economic theory seems in the light of experimental results to be more limited than expected. However, if the standard theory really does not apply to a vast portion of what has traditionally been considered the intended domain of economic theory, then economists have a *huge* problem to solve. Surely, the way to tackle the problem is by investigating the mechanisms that make the theory fail under certain experimental circumstances and work in others, which is exactly what those who first discovered the experimental anomalies of the standard theory were trying to do. The *we all know that it won't work in such conditions* attitude obscures an important point: the scope of economic theory is constantly redefined and worked out after a lot of empirical effort. The work of those who have devoted their careers to observing violations of the theory under a range of different "background" conditions has been and still is extremely useful and relevant. This research (Kahneman and Tversky's work is the explicit target of Binmore's dismissing remarks) highlights what the limitations and the domain of application of the theory are. Of course the anomalies do not *falsify* the theory – as some enemies of standard economic theory would like – for the simple reason that falsificationism is not the methodology of science (of *any* science, let alone experimental economics).

However, there is more than that. Binmore and others use the argument about the "proper" domain of economic theory to argue that the experiments that do not reproduce the conditions that make the standard models applicable (such as high incentives, repetition, etc.) somehow lack validity or interest. Should this position become widely accepted in the profession, experimentalists would be encouraged to take uncritically what in fact should be the very subject of their critical investigations – the mechanisms that make the theory work here but not there.

Conclusion: The context of experimentation

The controversy over economic incentives is a good playing field for someone interested in methodology. A commonly held assumption in the debate is that the adoption of a set of precepts or design recipes has multiple, perhaps universal, beneficial consequences. One of the themes of this book is that different experiments have different goals, and there are few (if any) universal recipes in experimental science. Whether a design is good or bad depends on what the research question is, or what hypothesis is under test. Similarly, external validity problems must be solved case by case; one cannot prove that a result is generalizable unless one specifies what the target system is. Quite obviously, economists are interested in several different phenomena, and therefore it is unlikely that a single experiment (or theory, for that matter) is a good tool for understanding the functioning of all of them at once.

We should keep in mind that *the experimental method itself* can sometimes be ill-suited to answer certain scientific questions. There is a "continuum" of methods going from "pure" laboratory experimentation to "social experiments" and "natural experiments" (Harrison and List, in press) in which the power to control variables (and hence to make reliable causal inferences) is usually traded against the ability to generalize from the specific system under study. One lesson of more mature sciences is that maximum strength and efficacy are achieved when *different* approaches are used to answer different questions, and all the various methods are intelligently combined in the course of a research program (think of the variety of approaches, from theoretical modeling to statistical simulation, biochemical experimentation, etc., that are used in biology).

Attempts to impose universal methodological standards, then, are likely to have a negative effect on the discipline. I do not mean to imply that methodologically speaking, experimental psychology, with its relatively more flexible standards, is in a healthier state than experimental economics or that psychologists have nothing to learn from economic practice. It is only natural that some exchange should take place between two neighbor disciplines, and the use of incentives in *some* contexts is one of the good practices economists have implemented successfully. Of course, incentives are a powerful tool – the important point is that they are not an all-purpose tool and must be used properly.

A more general point to be made at the end of this chapter (and of this book) is that it is dangerous to crystallize experimental practice into a set of rigid rules. Outsiders are often puzzled by the inflexible way in which economists implement incentive requirements while being extremely tolerant (if not slack) regarding other design issues – such as subject sampling, to name just one (Loewenstein 1999). Rules such as those embodied in the precepts are context specific and are justified only to the extent that they constitute low-level, concrete applications of more general methodological principles, such as those outlined in earlier chapters of this book. Because the problems faced by experimenters are diverse, it is unreasonable to impose a priori limitations at such a low methodological level – the rules are context specific, the context of experiments is varied and changes as new areas of economics are investigated experimentally. Eventually, what really matters is the experimenters' capacity to invent innovative designs to tackle important internal and external validity issues. We should not stipulate right from the start what the tools or the domain of experimental economics should be, because by so doing, we may hinder the most interesting research of the future.

APPENDIX A

The Karni-Safra Argument

According to Karni and Safra (1987), the BDM mechanism may be perceived by subjects as a two-stage lottery giving, among its outcomes, the possibility of playing out the priced gamble.[1] Suppose the gamble is $X = (4, 35/36; -1, 1/36)$ – one of the P-bets used by Grether and Plott, typically subject to preference reversals. If, by assumption, both $\pi(X)$ – the real selling price – and b – the bidding price – are restricted to the 1000 different values $0, 1/100, \ldots, k/100, \ldots, 9.99$ $(0 \leq k < 1000)$, the following two-stage lottery results from the BDM procedure:

$$
A = \left(\left(4, \frac{35}{36}; -1, \frac{1}{36} \right), \frac{\pi(X)}{10}; \delta_{\pi(X)}, \frac{1}{1000}; \delta_{\pi(X)+0.01}, \right.
$$
$$
\left. \frac{1}{1000}; \ldots; \delta_{9.99}, \frac{1}{1000} \right)
$$

where the δ_i stands for degenerate lotteries with probability 1 of getting i, and $\pi(X)/10$ is the probability of participating in X according to the BDM mechanism. Lottery A is equivalent to the tree in Figure A.1.

By definition of a certainty equivalent (CE), we know that $X \sim \delta_{CE(X)}$. Thus, by applying independence, there follows that

$$
A \sim A' = \left(CE \left(4, \frac{35}{36}; -1\frac{1}{36} \right), \frac{\pi(X)}{10}; \delta_{\pi(X)}, \frac{1}{1000}; \delta_{\pi(X)+0.01}, \right.
$$
$$
\left. \frac{1}{1000}; \ldots; \delta_{9.99}, \frac{1}{1000} \right).
$$

---

[1] I follow here the presentation given by Keller, Segal, and Wang (1993).

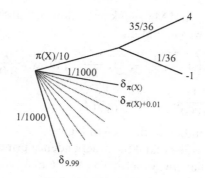

Figure A.1. Lottery A.

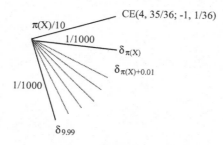

Figure A.2. Lottery A'.

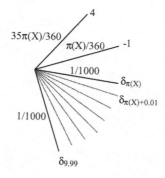

Figure A.3. Lottery R(A).

The indifference above implies that subjects see tree A.1 as equivalent to the tree in Figure A.2.

The task faced by a subject participating in a BDM experiment, then, can be represented as a maximization problem: what is the value of $\pi(X)$ that maximizes the value of lottery A'? An expected utility maximizer, as Becker, DeGroot, and Marschak have shown, will set $\pi(X) = CE(X)$. Now, by *reduction*, we can obtain

$$A \sim R(A) = \left(\left(4, \frac{35\pi(X)}{360}; -1, \frac{\pi(X)}{360} \right); \delta_{\pi(X)}, \frac{1}{1000}; \delta_{\pi(X)+0.01}, \right.$$
$$\left. \frac{1}{1000}; \ldots; \delta_{9.99}, \frac{1}{1000} \right)$$

with $R(A)$ corresponding to the tree in Figure A.3.

Karni and Safra argue that if the independence principle is *not* obeyed, then it is not true that *always* setting $\pi(X) = CE(X)$ maximizes the value of $R(A)$.

APPENDIX B

Subjective Bayesianism Again

Although in his recent papers, Vernon Smith does not say very much about external validity inferences, in an old article coauthored with Donald Rice, he supports a subjective Bayesian approach to this problem. Having rejected subjective Bayesianism as a solution to internal validity problems, unsurprisingly I do not find it palatable for external validity either. But let us see why in more detail.

Rice and Smith (1964) use the following concepts: H is a hypothesis under test.[1] N is a set of events or data from the "natural" or "real" world that confirm H, and E is a set of events or data collected in the laboratory that also confirm H. We assume, then, that both experimental and real-world evidence is available, and the problem is how to use them in combination in order to evaluate H. Unsurprisingly, subjective Bayesianism offers a solution to this problem, provided the appropriate prior probabilities are specified. What we are looking for is the posterior probability of H given E and N: $P(H \mid E \& N)$. According to Bayes's theorem,

$$P(H \mid E \& N) = \frac{P(E \& N \mid H) P(H)}{P(E \& N)}.$$

In order to solve the equation, we need to specify the priors $P(H)$, $P(E \& N \mid H)$, and $P(E \& N)$. Using the principle of total probability, $P(E \& N)$ can be derived from $P(E \& N \mid H) \times P(H)$ and $P(E \& N \mid \sim H) \times P(\sim H)$. Therefore, all we need is a consistent probability assignment to

---

[1] According to Rice and Smith (1964), H is normally derived from a theoretical model; as I say repeatedly in the first part of the book, I do not think this is generally true of experimental hypotheses, but it does not matter for the present discussion.

$P(H)$, $P(\sim H)$, $P(E \;\&\; N \mid H)$, and $P(E \;\&\; N \mid \sim H)$. If we could then also extend the assignment to $P(\sim E \;\&\; N \mid H)$, $P(E \;\&\; \sim N \mid H)$, $P(\sim E \;\&\; \sim N \mid H)$, $P(\sim E \;\&\; N \mid \sim H)$, $P(E \;\&\; \sim N \mid \sim H)$, and $P(\sim E \;\&\; \sim N \mid \sim H)$, we would be able to calculate the posteriors $P(H \mid \sim E \;\&\; N)$, $P(H \mid E \;\&\; \sim N)$, and $P(H \mid \sim E \;\&\; \sim N)$. Every possible combination of (positive or negative) field and (positive or negative) laboratory evidence could be taken into account. As Rice and Smith point out,

> [the conditional priors] express the scientist's a priori degree of belief in the reliability of the observations in the laboratory and in the natural world; they also express his degree of belief in the relevance of each kind of evidence to the hypothesis under study. (Rice and Smith 1964, p. 60)

The main problem with this "solution" to the external validity challenge is the usual one with subjective Bayesianism. It seems wrong to let individual scientists' prejudice affect the evidential relation decisively, without imposing any normative requirements on the formation of priors. A scientist who is dogmatically opposed to the use of laboratory evidence for the assessment of economic hypotheses, for example, can always set his priors so as to make the impact of E on H equal to nothing. Rice and Smith simply note that the above-mentioned scientist will be careful to set his prior beliefs so that $P(E \;\&\; N \mid H) = P(\sim E \;\&\; N \mid H)$ and $P(E \;\&\; \sim N \mid H) = P(\sim E \;\&\; \sim N \mid H)$. However, the interesting question is, Would such an assignment be *reasonable*? Surely Rice and Smith would not be happy if the whole of the economic profession assigned zero weight to *all* empirical evidence collected in the laboratory – but then what is the difference between "convincing" and "unconvincing" evidence? What is it that makes one's priors (assuming we have any such prior beliefs at all, that is) reasonable?

Until one is able to answer these questions, the problem of external validity has not been properly solved. Subjective Bayesianism provides only a way of framing the problem in a formally elegant way but provides no genuine solution to the interesting methodological questions. One of my teachers at graduate school, Elie Zahar, used to say that "subjective Bayesianism accommodates everything and explains nothing." In my view, he was quite right. We need to go beyond the Bayesian account in order to achieve a genuine understanding of the external validity problem.

Bibliography

Achinstein, P. (2001) *The Book of Evidence*. Oxford: Oxford University Press.

Ackermann, R. (1989) "The New Experimentalism," *British Journal for the Philosophy of Science* 40: 185–90.

Allais, M. (1953) "The Foundations of a Positive Theory of Choice Involving Risk and a Criticism of the Postulate and Axioms of the American School," in M. Allais and O. Hagen (eds. 1979) *Expected Utility Hypothesis and the Allais Paradox*. Dordrecht: Reidel, pp. 257–332.

American Economic Review (1992) Symposium on "Theory and Misbehavior of First-price Auctions." Vol. 82, no. 5, pp. 1374–443.

Anand, P. (1993) *Foundations of Rational Choice under Risk*. Oxford: Oxford University Press.

Anderson, S. P., J. K. Goeree, and C. A. Holt (1998) "A Theoretical Analysis of Altruism and Decision Error in Public Goods Games," *Journal of Public Economics* 70: 297–323.

Andreoni, J. (1995) "Warm Glow vs. Cold Prickle: The Effects of Positive and Negative Framing on Cooperation in Experiments," *Quarterly Journal of Economics* 110: 1–21.

Ankeny, R. (2001) "Model Organisms as Models: Understanding the 'Lingua Franca' of the Human Genome Project," *Philosophy of Science* 68: S251–61.

Anscombe, E. (1971) *Causality and Determinism*. Cambridge: Cambridge University Press.

Aristotle. *Topics Books I and VII, with Excerpts from Related Texts*. Oxford: Clarendon, 1997.

Ayres, I. and P. Cramton (1996) "Deficit Reduction through Diversity: A Case Study of How Affirmative Action at the FCC Increased Auction Competition," *Stanford Law Review* 48: 761–815.

Backhouse, R. E. (1994) "The Lakatosian Legacy in Economic Methodology," in R. E. Backhouse (ed.) *New Directions in Economic Methodology*. London: Routledge, pp. 173–91.

Backhouse, R. E. (1997) *Truth and Progress in Economic Knowledge.* Cheltenham: Edward Elgar.

Bacon, F. (1620) *Novum Organum.* Chicago: Open Court, 1994.

Bardsley, N. (in press) "Experimental Economics and the 'Artificiality of Alteration,'" *Journal of Economic Methodology.*

Baron, J. (2001) "Purposes and Method," *Behavioral and Brain Sciences* 24: 403.

Becker, G. M., M. H. DeGroot, and J. Marschak (1963) "Stochastic Models of Choice Behaviour," *Behavioral Science* 8: 41–55.

Becker, G. M., M. H. DeGroot, and J. Marschak (1964) "Measuring Utility by a Single-Response Sequential Method," *Behavioral Science* 9: 226–32.

Berg, J. E., J. W. Dickhaut, and J. R. O'Brien (1985) "Preference Reversal and Arbitrage," *Research in Experimental Economics* 3: 31–71.

Bergstrom, T. C. and J. H. Miller (1997) *Experiments with Economic Principles: Microeconomics.* New York: McGraw-Hill.

Berkovitz, L. and E. Donnerstein (1982) "External Validity Is More than Skin Deep," *American Psychologist* 37: 245–57.

Bernard, C. (1865) *Introduction à l'étude de la Medicine Experimentale.* Paris: Flammarion; Engl. transl. *Introduction to the Study of Experimental Medicine.* New York: Henri Schumann, 1957.

Betsch, T. and S. Haberstroh (2001) "Financial Incentives Do Not Pave the Road to Good Experimentation," *Behavioral and Brain Sciences* 24: 404.

Binmore, K. (1992) *Fun and Games: A Text on Game Theory.* Lexington, Mass.: D. C. Heath & Co.

Binmore, K. (1999) "Why Experiment in Economics?," *Economic Journal* 109: F16–24.

Binmore, K. and P. Klemperer (2002) "The Biggest Auction Ever: The Sale of the British 3G Telecom Licences," *Economic Journal* 112: C74–96.

Blaug, M. (1980) *The Methodology of Economics.* Cambridge: Cambridge University Press.

Bogen, J. and J. Woodward (1988) "Saving the Phenomena," *Philosophical Review* 97: 303–52.

Boumans, M. and M. S. Morgan (2000) "*Ceteris Paribus* Conditions: Materiality and the Application of Economic Theory," *Journal of Economic Methodology* 8: 11–26.

Brown, J. R. (1991) *Laboratory of the Mind: Thought Experiments in the Natural Sciences.* London: Routledge.

Brown, J. R. (1994) *Smoke and Mirrors: How Science Reflects Reality.* London: Routledge.

Brunsvik, E. (1955) "Representative Design and Probabilistic Theory in a Functional Psychology," *Psychological Review* 62: 193–217.

Buchwald, J. Z. (1994) *The Creation of Scientific Effects.* Chicago: University of Chicago Press.

Burlando, R. M. and F. Guala (in press) "Heterogeneous Agents in Public Goods Experiments," *Experimental Economics.*

Burlando, R. M. and P. Webley (1999) "Individual Differences and Long-run Equilibria in a Public Good Experiment," in *Inquiries into the Nature and Causes of Behavior*. Proceedings of the 24th IAREP Annual Colloquium, Belgirate, Italy.

Bykowsky, M. M., R. J. Cull, and J. O. Ledyard (2000) "Mutually Destructive Bidding: The FCC Auction Design Problem," *Journal of Regulatory Economics* 17: 205–28.

Caldwell, B. J. (1991) "Clarifying Popper," *Journal of Economic Literature* 29: 1–33.

Camerer, C. F. (1995) "Individual Decision Making," in J. H. Kagel and A. E. Roth (eds.) *The Handbook of Experimental Economics*. Princeton: Princeton University Press, pp. 587–703.

Camerer, C. F. and R. M. Hogarth (1999) "The Effects of Financial Incentives in Experiments: A Review and Capital-Labor-Production Framework," *Journal of Risk and Uncertainty* 19: 7–42.

Camerer, C. F., G. Loewenstein, and E. D. Prelec (in press) "Neuroeconomics: How Neuroscience Can Inform Economics," *Journal of Economic Perspectives*.

Campbell, C. M., J. H. Kagel, and D. Levin (1999) "The Winner's Curse and Public Information in Common Value Auctions: Reply," *American Economic Review* 89: 325–34.

Campbell, D. and J. Stanley (1963) *Experimental and Quasi–Experimental Designs for Research*. Chicago: Rand McNally.

Capen, E. C., R. V. Clapp, and W. M. Campbell (1971) "Competitive Bidding in High-Risk Situations," *Journal of Petroleum Technology* 23: 641–53.

Carnap, R. (1950) *Logical Foundations of Probability*. Chicago: University of Chicago Press.

Carpenter, K. J. (1986) *The History of Scurvy and Vitamin C*. Cambridge: Cambridge University Press.

Cartwright, N. (1983) *How the Laws of Physics Lie*. Oxford: Clarendon Press.

Cartwright, N. (1989) *Nature's Capacities and Their Measurement*. Oxford: Oxford University Press.

Cartwright, N. (1991) "Replicability, Reproducibility, and Robustness: Comments on Harry Collins," *History of Political Economy* 23: 143–55.

Cartwright, N. (1999) *The Dappled World: A Study of the Boundaries of Science*. Cambridge: Cambridge University Press.

Chew, S. H. and K. MacCrimmon (1979) "Alpha-Nu Choice Theory: A Generalization of Expected Utility Theory," Working Paper no. 686. Faculty of Commerce and Business Administration, University of British Columbia.

Christensen, L. B. (2001) *Experimental Methodology*, 8th ed. Needham Heights, Mass.: Allyn & Bacon.

Chu, Y. P. and R. L. Chu (1990) "The Subsidence of Preference Reversals in Simplified and Marketlike Experimental Settings: A Note," *American Economic Review* 80: 902–11.

Collins, H. M. (1985) *Changing Order: Replication and Induction in Scientific Practice*. London: Sage.

Collins, H. M. (1994) "A Strong Confirmation of the Experimenter's Regress," *Studies in the History and Philosophy of Science* 25: 493–503.

Collins, H. M. and T. Pinch (1993) *The Golem: What You Should Know about Science*. Cambridge: Cambridge University Press.

Conlisk, J. (1983) "Three Variants on the Allais Example," *American Economic Review* 79: 392–407.

Cook, T. and D. Campbell (1979) *Quasi–experimentation: Design and Analysis Issues for Field Settings*. Chicago: Rand McNally.

Cox, J. C., S. H. Dinkin, and V. L. Smith (1999) "The Winner's Curse and Public Information in Common Value Auctions: Comment," *American Economic Review* 89: 319–24.

Cox, J. C. and S. Epstein (1989) "Preference Reversals without the Independence Axiom," *American Economic Review* 79: 408–26.

Cox, J. C. and D. M. Grether (1996) "The Preference Reversal Phenomenon: Response Mode, Markets and Incentives," *Economic Theory* 7: 381–405.

Cox, J. C. and R. M. Isaac (1986) "Experimental Economics and Experimental Psychology: Ever the Twain Shall Meet?," in A. J. MacFadyen and H. W. MacFadyen (eds.) *Economic Psychology: Interactions in Theory and Application*. New York: North Holland, pp. 647–69.

Cox, J. C., B. Robertson, and V. L. Smith (1982) "Theory and Behavior of Single Object Auctions," in V. L. Smith (ed.) *Research in Experimental Economics*. Greenwich, Conn.: JAI Press, pp. 1–43.

Cramton, P. C. (1995) "Money Out of Thin Air: The Nationwide Narrowband PCS Auction," *Journal of Economics and Management Strategy* 4: 267–343.

Cramton, P. C. (1997) "The FCC Spectrum Auctions: An Early Assessment," *Journal of Economics and Management Strategy* 6: 431–95.

Cramton, P. C. (1998) "The Efficiency of the FCC Spectrum Auctions," *Journal of Law and Economics* 41: 727–36.

Cramton, P. C. and J. Schwartz (2000) "Collusive Bidding: Lessons from the FCC Spectrum Auction," *Journal of Regulatory Economics* 17: 229–52.

Cross, R. (1982) "The Duhem-Quine Thesis, Lakatos and the Appraisal of Theories in Macroeconomics," *Economic Journal* 92: 320–40.

Cubitt, R. P. (in press) "Experiments and the Domain of Economic Theory," *Journal of Economic Methodology*.

Cubitt, R. P., C. Starmer, and R. Sugden (2001) "Discovered Preferences and the Experimental Evidence of Violations of Expected Utility Theory," *Journal of Economic Methodology* 8: 385–414.

Cunningham, A. and P. Williams (eds. 1992) *The Laboratory Revolution in Medicine*. Cambridge: Cambridge University Press.

Davis, D. D. and C. H. Holt (1993) *Experimental Economics*. Princeton: Princeton University Press.

Dawes, R. M. and Thaler, R. H. (1988) "Anomalies: Cooperation," *Journal of Economic Perspectives* 2: 187–97.

de Vroey, M. (1998) "Is the Tâtonnement Hypothesis a Good Caricature of Market Forces?," *Journal of Economic Methodology* 5: 201–22.

Dooley, D. (2001) *Social Research Methods*, 3rd ed. London: Prentice Hall.

Dorling, J. (1979) "Bayesian Personalism, the Methodology of Research Programmes, and Duhem's Problem," *Studies in History and Philosophy of Science* 10: 177–87.

Duhem, P. (1906) *La théorie physique. Son objet et sa structure.* Paris: Chevalier et Rivière; Engl. transl. *The Aim and Structure of Physical Theory.* Princeton: Princeton University Press, 1954.

Dupré, J. (1984) "Probabilistic Causality Emancipated," *Midwest Studies in Philosophy* 9: 169–75.

Dupré, J. (1993) *The Disorder of Things.* Cambridge, Mass.: Harvard University Press.

Dupré, J. (2001) "Economics without Mechanism," in U. Mäki (ed.) *The Economic World View.* Cambridge: Cambridge University Press, pp. 308–32.

Earman, J. (1992) *Bayes or Bust? A Critical Examination of Bayesian Confirmation Theory.* Cambridge, Mass.: MIT Press.

Earman, J. and C. Glymour (1980) "Relativity and the Eclipses: The British Eclipse Expedition and Their Predecessors," *Historical Studies in the Physical Sciences* 11: 49–85.

Economics Focus (1999) "News from the Lab," *The Economist*, May 8, p. 96.

Ehrhart, K. M. and C. Keser (1999) "Mobility and Cooperation: On the Run," Working Paper 99s–24, CIRANO, University of Montreal.

Falk, A. and U. Fischbacher (2000) "A Theory of Reciprocity," Working Paper 6/2000. Institute for Empirical Research in Economics, University of Zurich.

Fehr, E. and U. Fishbacher (2002) "Why Social Preferences Matter – The Impact of Non-selfish Motives on Competition, Cooperation and Incentives," *Economic Journal* 112: C1–33.

Feyerabend, P. K. (1975) *Against Method.* London: Verso, 2nd ed. 1993.

Fischbacher, U., S. Gächter, and E. Fehr (2001) "Are People Conditionally Cooperative? Evidence from a Public Goods Experiment," *Economics Letters* 71: 397–404.

Fisher, R. A. (1956) *Statistical Methods and Scientific Inference.* Edinburgh: Oliver and Boyd.

Fischhoff, B. (1982) "Debiasing," in D. Kahneman, P. Slovic, and A. Tversky (eds.) *Judgment under Uncertainty.* Cambridge: Cambridge University Press, pp. 422–44.

Fischhoff, B. (1996) "The Real World: What Good Is It?," *Organizational Behavior and Human Decision Processes* 65: 232–48.

Flexner, S. and P. Lewis (1910) "Experimental Poliomyelitis in Monkeys; Active Immunization and Passive Serum Protection," *Journal of the American Medical Association* 54: 1780.

Fodor, J. A. (1974) "Special Sciences (or: The Disunity of Science as a Working Hypothesis)," *Synthese* 28: 97–115.

Fodor, J. A. (1987) *Psychosemantics.* Cambridge, Mass.: MIT Press.

Fodor, J. A. (1989) "Making Mind Matter More," *Philosophical Topics* 17: 59–79.

Fontaine, P. (2002) "Blood, Politics, and Social Science," *Isis* 94: 401–34.

Forster, M. and E. Sober (1994) "How to Tell When Simpler, More Unified, or Less *Ad Hoc* Theories Will Provide More Accurate Predictions," *British Journal for the Philosophy of Science* 45: 1–35.

Forster, M. and E. Sober (in press) "Why Likelihood?," in M. Taper and S. Lee (eds.) *The Nature of Scientific Evidence.* Chicago: University of Chicago Press.

Fouraker, L. E. and S. Siegel (1963) *Bargaining Behavior*. New York: McGraw-Hill.

Frank, R., T. Gilovich, and D. Regan (1993) "Does Studying Economics Inhibit Cooperation?," *Journal of Economic Perspectives* 7: 159–71.

Frankfort-Nachmias, C. and D. Nachmias (1996) *Research Methods in the Social Sciences*. London: Arnold.

Franklin, A. (1986) *The Neglect of Experiment*. Cambridge: Cambridge University Press.

Franklin, A. (1990) *Experiment, Right or Wrong*. Cambridge: Cambridge University Press.

Franklin, A. (1994) "How to Avoid the Experimenter's Regress," *Studies in History and Philosophy of Science* 15: 51–62.

Franklin, A. (1998) "Experiment in Physics," in E. N. Zalta (ed.) *The Stanford Encyclopaedia of Philosophy*, http://plato.stanford.edu/entries/physics-experiment.

Frey, B. and S. Meier (2003) "Are Political Economists Selfish and Indoctrinated? Evidence from a Natural Experiment," *Economic Inquiry* 41: 448–62.

Friedman, D. and A. Cassar (2004) *Economics Lab: An Intensive Course in Experimental Economics*. London: Routledge.

Friedman, D. and S. Sunder (1994) *Experimental Methods: A Primer for Economists*. Cambridge: Cambridge University Press.

Friedman, M. (1953) "The Methodology of Positive Economics," in *Essays in Positive Economics*. Chicago: University of Chicago Press, pp. 3–43.

Friedman, M. (1999) *Reconsidering Logical Positivism*. Cambridge: Cambridge University Press.

Gachter, S. and C. Thoni (2004) "Social Learning and Voluntary Cooperation among Like-Minded People," unpublished paper, University of St. Gallen.

Galison, P. (1987) *How Experiments End*. Chicago: University of Chicago Press.

Galison, P. (1997) *Image and Logic*. Chicago: University of Chicago Press.

Giere, R. N. (1977) "Testing vs. Information Models of Scientific Inference," in R. G. Colodny (ed.) *Logic, Laws, and Life: Some Philosophical Complications*. University of Pittsburgh Series in the Philosophy of Science, Vol. 6. Pittsburgh: University of Pittsburgh Press, pp. 19–70.

Giere, R. N. (1979) *Understanding Scientific Reasoning*. New York: Harcourt Brace, 4th ed. 1997.

Giere, R. N. (1983) "Testing Theoretical Hypotheses," in J. Earman (ed.) *Testing Scientific Theories*. Minnesota Studies in the Philosophy of Science, Vol. 10. Minneapolis: University of Minnesota Press, pp. 269–98.

Giere, R. N. (1988) *Explaining Science*. Chicago: University of Chicago Press.

Giere, R. N. (2002) "Models as Parts of Distributed Cognitive Systems," in L. Magnani and N. J. Nersessian (eds.) *Model-Based Reasoning: Science, Technology, Values*. New York: Kluwer, pp. 227–41.

Gigerenzer, G., P. Todd, and the ABC Research Group (1999) *Simple Heuristics that Make Us Smart*. Oxford: Oxford University Press.

Gillies, D. (1991) "Intersubjective Probability and Confirmation Theory," *British Journal for the Philosophy of Science* 42: 513–33.

Gillies, D. (1993) *Philosophy of Science in the Twentieth Century*. Oxford: Blackwell.

Glennan, S. (2002) "Rethinking Mechanistic Explanation," *Philosophy of Science* 69: S342–53.

Gooding, D. (1990) *Experiment and the Making of Meaning*. Dordrecht: Kluwer.

Granger, C. W. J. (1980) "Testing for Causality: A Personal Viewpoint," *Journal of Economic Dynamics and Control* 2: 329–52.

Grether, D. and C. Plott (1979) "Economic Theory of Choice and the Preference Reversal Phenomenon," *American Economic Review* 69: 623–38.

Guala, F. (1998) "Experiments as Mediators in the Non-Laboratory Sciences," *Philosophica* 62: 901–18.

Guala, F. (1999) "The Problem of External Validity (or 'Parallelism') in Experimental Economics," *Social Science Information* 38: 555–73.

Guala, F. (2000a) "Artefacts in Experimental Economics: Preference Reversals and the Becker-DeGroot-Marschak Mechanism," *Economics and Philosophy* 16: 47–75.

Guala, F. (2000b) "The Logic of Normative Falsification: Rationality and Experiments in Decision Theory," *Journal of Economic Methodology* 7: 59–93.

Guala, F. (2001) "Building Economic Machines: The FCC Auctions," *Studies in History and Philosophy of Science* 32: 453–77.

Guala, F. (2002a) "Models, Simulations, and Experiments," in L. Magnani and N. J. Nersessian (eds.) *Model-Based Reasoning: Science, Technology, Values*. New York: Kluwer, pp. 59–74.

Guala, F. (2002b) "On the Scope of Experiments in Economics: Comments on Siakantaris," *Cambridge Journal of Economics* 26: 261–7.

Guala, F. (2003) "Experimental Localism and External Validity," *Philosophy of Science* 70: 1195–205.

Guala, F. and L. Mittone (2002) "Experiments in Economics: Testing Theories vs. the Robustness of Phenomena," CEEL Working Paper 09–02, University of Trento.

Hacking, I. (1965) *Logic of Statistical Inference*. Cambridge: Cambridge University Press.

Hacking, I. (1983) *Representing and Intervening*. Cambridge: Cambridge University Press.

Hacking, I. (1988) "The Participant Irrealist at Large in the Laboratory," *British Journal for the Philosophy of Science* 39: 277–94.

Hacking, I. (1989) "Extragalactic Reality: The Case of Gravitational Lensing," *Philosophy of Science* 56: 555–81.

Hacking, I. (1992) "The Self-Vindication of the Laboratory Sciences," in A. Pickering (ed.) *Science as Practice and Culture*. Chicago: University of Chicago Press, pp. 29–64.

Hands, D. W. (1985) "Second Thoughts on Lakatos," *History of Political Economy* 17: 1–16.

Hands, D. W. (2001) *Reflection without Rules: Economic Methodology and Contemporary Science Theory*. Cambridge: Cambridge University Press.

Harding, S. (ed. 1976) *Can Theories Be Refuted?* Dordrecht: Reidel.

Hargreaves Heap, S. and Y. Varoufakis (1995) "Experimenting with Neoclassical Economics: A Critical Review of Experimental Economics," in I. H. Rima (ed.) *Measurement, Quantification and Economic Analysis.* London: Routledge.

Harrison, G. W. (1989) "Theory and Misbehavior of First-Price Auctions," *American Economic Review* 79: 749–62.

Harrison, G. W. (1994) "Expected Utility Theory and the Experimentalists," *Empirical Economics* 19: 223–54.

Harrison, G. W. and J. A. List (2004) "Field Experiments," *Journal of Economic Literature* 42(4), 1013–59.

Hausman, D. M. (1989) "Ceteris Paribus Clauses and Causality in Economics," *PSA 1988*, Vol. 2. East Lansing: Philosophy of Science Association.

Hausman, D. M. (1990) "Supply and Demand Explanations and Their Ceteris Paribus Clauses," *Review of Political Economy* 2: 168–87; reprinted in *Essays on Philosophy and Economic Methodology.* Cambridge: Cambridge University Press, 1992.

Hausman, D. M. (1992a) *The Inexact and Separate Science of Economics.* Cambridge: Cambridge University Press.

Hausman, D. M. (1992b) "Why Look under the Hood?," in *Essays in Philosophy and Economic Methodology.* Cambridge: Cambridge University Press, pp. 70–3.

Hausman, D. M. (1998a) *Causal Asymmetries.* Cambridge: Cambridge University Press.

Hausman, D. M. (1998b) "Problems with Realism in Economics," *Economics and Philosophy* 14: 185–213.

Hausman, D. M. (2000) "Revealed Preference, Belief, and Game Theory," *Economics and Philosophy* 16: 99–115.

Hausman, D. M. (2001) "Explanation and Diagnosis in Economics," *Revue internationale de philosophie* 217: 311–26.

Hausman, D. M. (in press) "Constructing Experimental Games," *Journal of Economic Methodology.*

Hausman, D. M. (unpublished) "Probabilistic Causality and Practical Causal Generalizations," University of Wisconsin-Madison.

Hausman, D. M. and P. Mongin (1998) "Economists' Responses to Anomalies: Full-Cost Pricing versus Preference Reversals," in J. Davis (ed.) *New Economics and Its History.* History of Political Economy Supplement, Vol. 29. Durham: Duke University Press, pp. 255–72.

Hempel, C. G. (1952) *Fundamentals of Concept-Formation in Empirical Science.* Chicago: University of Chicago Press.

Hempel, C. G. (1965) *Aspects of Scientific Explanation.* New York: Free Press.

Hendry, R. (1980) "Econometrics – Alchemy or Science?," *Economica* 47: 387–406.

Henshel, R. L. (1980) "The Purposes of Laboratory Experimentation and the Virtues of Deliberate Artificiality," *Journal of Experimental Social Psychology* 16: 466–78.

Herstein, I. and J. Milnor (1953) "An Axiomatic Approach to Measurable Utility," *Econometrica* 47: 291–7.

Hertwig, R. and A. Ortmann (2001) "Experimental Practices in Economics: A Methodological Challenge for Psychologists?," *Behavioral and Brain Sciences* 24: 383–451.

Hesse, M. B. (1963) *Models and Analogies in Science*. London: Sheed & Ward.

Hey, J. D. (1991) *Experiments in Economics*. Oxford: Blackwell.

Hogarth, R. M. and M. W. Reder (eds. 1986) *Rational Choice: The Contrast between Economics and Psychology*. Chicago: University of Chicago Press.

Holst, A. and T. Frölich (1907) "Experimental Studies Relating to Ship-Beri-Beri and Scurvy. II., On the Etiology of Scurvy," *Journal of Hygiene* 7: 634–71.

Holt, C. A. (1986) "Preference Reversals and the Independence Axiom," *American Economic Review* 76: 508–15.

Hon, G. (1989) "Towards a Typology of Experimental Error: An Epistemological View," *Studies in History and Philosophy of Science* 20: 469–504.

Hoover, K. D. (2001) *Causality in Macroeconomics*. Cambridge: Cambridge University Press.

Howson, C. (1997a) "A Logic of Induction," *Philosophy of Science* 64: 268–90.

Howson, C. (1997b) "Error Probabilities in Error," *Philosophy of Science* 64: S185–94.

Howson, C. and P. Urbach (1989) *Scientific Reasoning: The Bayesian Approach*. Chicago: Open Court.

Hughes, R. I. G. (1997) "Models and Representation," *Philosophy of Science* 64: S325–36.

Hughes, R. I. G. (1999) "The Ising Model, Computer Simulation, and Universal Physics," in M. S. Morgan and M. C. Morrison (eds.) *Models as Mediators*. Cambridge: Cambridge University Press, pp. 97–145.

Hume, D. (1740) *A Treatise of Human Nature*. Oxford: Clarendon Press, 1978.

Humphreys, P. (1989) *The Chances of Explanation*. Princeton: Princeton University Press.

Ingrao, B. and G. Israel (1987) *La mano invisibile*. Bari: Laterza; Engl. transl. *The Invisible Hand*. Cambridge, Mass.: MIT Press, 1990.

Isaac, R. M., K. F. McCue, and C. R. Plott (1985) "Public Goods Provision in an Experimental Environment," *Journal of Public Economics* 26: 51–74.

Isaac, R. M. and J. M. Walker (1998) "Nash as an Organizing Principle in the Voluntary Provision of Public Goods: An Experimental Analysis," *Experimental Economics* 1: 191–206.

Isaac, R. M., J. M. Walker, and S. Thomas (1984) "Divergent Evidence on Free-Riding: An Experimental Examination of Possible Explanations," *Public Choice* 43: 113–49.

Kagel, J. H. and D. Levin (1986) "The Winner's Curse Phenomenon and Public Information in Common Value Auctions," *American Economic Review* 76: 894–920.

Kagel, J. H. and A. E. Roth (eds. 1995) *The Handbook of Experimental Economics*. Princeton: Princeton University Press.

Karni, E. and Z. Safra (1987) "'Preference Reversal' and the Observability of Preferences by Experimental Methods," *Econometrica* 55: 675–85.

Keller, L. R., U. Segal, and T. Wang (1993) "The Becker-DeGroot-Marschak Mechanism and Generalized Utility Theories: Theoretical Predictions and Empirical Observations," *Theory and Decision* 34: 83–97.

Keser, C. (1996) "Voluntary Contributions to a Public Good When Partial Contribution Is a Dominant Strategy," *Economics Letters* 50: 359–66.

Kim, O. and J. M. Walker (1984) "The Free Rider Problem: Experimental Evidence," *Public Choice* 43: 3–24.

Kincaid, H. (1996) *Philosophical Foundations of the Social Sciences*. Cambridge: Cambridge University Press.

Kitcher, P. (1981) "Explanatory Unification," *Philosophy of Science* 48: 507–31.

Kitcher, P. (1993) *The Advancement of Science*. Oxford: Oxford University Press.

Klemperer, P. (2002) "How (Not) to Run Auctions: The European 3G Telecom Auctions," *European Economic Review* 46: 829–45.

Klemperer, P. (2004) *Auctions: Theory and Practice*. Princeton: Princeton University Press.

Knez, M. and V. L. Smith (1987) "Hypothetical Valuations and Preference Reversals in the Context of Asset Trading," in A. E. Roth (ed.) *Laboratory Experimentation in Economics: Six Points of View*. Cambridge: Cambridge University Press, pp. 131–54.

Kohler, R. E. (1994) *Lords of the Fly*. Chicago: University of Chicago Press.

Kruglanski, A. W. (1975) "The Human Subject in the Psychology Experiment: Fact and Artifact," in L. Berkovitz (ed.) *Advances in Experimental Social Psychology*, Vol. 8. New York: Academic Press, pp. 101–47.

Kuhn, T. S. (1962) *The Structure of Scientific Revolutions*. Chicago: University of Chicago Press, 2nd ed. 1970.

Kwerel, E. R. (2004) "Foreword," in P. Milgrom, *Putting Auction Theory to Work*. Cambridge: Cambridge University Press, pp. xv–xxii.

Kwerel, E. R. and G. L. Rosston (2000) "An Insider's View of FCC Spectrum Auctions," *Journal of Regulatory Economics* 17: 253–89.

LaFollette, H. and N. Shanks (1995) "Two Models of Models in Biomedical Research," *Philosophical Quarterly* 45: 141–60.

Lakatos, I. (1970) "Falsificationism and the Methodology of Scientific Research Programmes," in *The Methodology of Scientific Research Programmes*. Philosophical Papers, Vol. 1. Cambridge: Cambridge University Press, 1978, pp. 8–101.

Lakatos, I. (1974) "Popper on Demarcation and Induction," in *The Methodology of Scientific Research Programmes*. Philosophical Papers, Vol. 1. Cambridge: Cambridge University Press, 1978, pp. 139–67.

Latour, B. (1984) *Les microbes: guerre et paix*. Paris: Métailié; Engl. transl. *The Pasteurisation of France*. Cambridge, Mass.: Harvard University Press, 1988.

Latour, B. (1987) *Science in Action*. Cambridge, Mass.: Harvard University Press.

Latour, B. (1988) "Comments on 'The Sociology of Knowledge of Child Abuse'," *Nous* 22: 67–9.

Latour, B. and S. Woolgar (1979) *Laboratory Life: The Construction of Scientific Facts*. Princeton: Princeton University Press, 2nd ed. 1986.

Latsis, S. (ed. 1976) *Method and Appraisal in Economics*. Cambridge: Cambridge University Press.

Lawson, T. (1997) *Economics and Reality*. London: Routledge.

Ledyard, J. O. (1995) "Public Goods: A Survey of Experimental Research," in J. H. Kagel and A. E. Roth (eds.) *The Handbook of Experimental Economics*. Princeton: Princeton University Press, pp. 111–94.

Ledyard, J. O., D. Porter, and A. Rangel (1997) "Experiments Testing Multiobject Allocation Mechanisms," *Journal of Economics and Management Strategy* 6: 639–75.

Lee, K. S. (unpublished) "Rationality, Minds, and Machines in the Laboratory: A Thematic History of Vernon Smith's Experimental Economics," Ph.D. dissertation, University of Notre Dame.

Leonard, R. (1994) "Laboratory Strife: Higgling as Experimental Science in Economics and Social Psychology," in N. B. De Marchi and M. S. Morgan (eds.) *Higgling*. History of Political Economy Supplement, Vol. 26. Durham: Duke University Press.

Lichtenstein, S. and P. Slovic (1968) "Relative Importance of Probabilities and Payoffs in Risk-Taking," *Journal of Experimental Psychology* Supplement, Part 2: 1–18.

Lichtenstein, S. and P. Slovic (1971) "Reversals of Preference Between Bids and Choices in Gambling Decisions," *Journal of Experimental Psychology* 89: 46–55.

Lichtenstein, S. and P. Slovic (1973) "Response-Induced Reversals of Preference in Gambling: An Extended Replication in Las Vegas," *Journal of Experimental Psychology* 101: 16–20.

Lipton, P. (1991) *Inference to the Best Explanation*. London: Routledge.

Loewenstein, G. (1999) "Experimental Economics from the Vantage-Point of Behavioral Economics," *Economic Journal* 109: F25–34.

Loomes, G. (1989) "Experimental Economics," in J. D. Hey (ed.) *Current Issues in Microeconomics*. New York: St. Martin's Press, pp. 152–78.

Loomes, G., C. Starmer, and R. Sugden (1989) "Preference Reversal: Information-Processing Effect or Rational Non-Transitive Choice?," *Economic Journal* 99: 140–51.

Loomes, G. and R. Sugden (1995) "Incorporating a Stochastic Element into Decision Theories," *European Economic Review* 39: 641–8.

Lucas, R. E. (1982) *Studies in Business Cycle Theory*. Cambridge, Mass.: MIT Press.

Luce, R. D. and H. Raiffa (1957) *Games and Decisions*. New York: Wiley.

Lynch, M. (1985) *Art and Artifact in Laboratory Science*. London: Routledge.

Machamer, P., L. Darden, and C. F. Craver (2000) "Thinking about Mechanisms," *Philosophy of Science* 67: 1–25.

Machina, M. J. (1982) "'Expected Utility' Analysis without the Independence Axiom," *Econometrica* 50: 277–323.

Mackie, J. L. (1974) *The Cement of the Universe*. Oxford: Clarendon Press.

Mäki, U. (1996) "Scientific Realism and Some Peculiarities of Economics," *Boston Studies in the Philosophy of Science* 69: 424–65.

Mäki, U. (2001a) "Models," in N. J. Smelser and P. B. Baltes (eds.) *The International Encyclopaedia of Social and Behavioral Sciences*, Vol. 15. London: Elsevier, pp. 9931–7.

Mäki, U. (2001b) "Explanatory Unification: Double and Doubtful," *Philosophy of the Social Sciences* 31: 488–506.

Mäki, U. (in press) "Models Are Experiments, Experiments Are Models," *Journal of Economic Methodology*.

Mäki, U. and J. P. Piimies (1998) "Ceteris Paribus," in J. B. Davis, D. W. Hands, and U. Mäki (eds.) *The Handbook of Economic Methodology*. Cheltenham: Elgar, pp. 55–9.

Marschak, J. (1950) "Rational Behaviour, Uncertain Prospects, and Measurable Utility," *Econometrica* 18: 111–41.

Marwell, G. and R. E. Ames (1981) "Economists Free Ride, Does Anyone Else?," *Journal of Public Economics* 15: 295–310.

Mas-Colell, A., M. D. Whinston, and J. R. Green (1995) *Microeconomic Theory.* Oxford: Oxford University Press.

Mayo, D. (1996) *Error and the Growth of Experimental Knowledge.* Chicago: University of Chicago Press.

Mayo, D. (1997a) "Response to Howson and Laudan," *Philosophy of Science* 64: 323–33.

Mayo, D. (1997b) "Error Statistics and Learning from Error: Making a Virtue of Necessity," *Philosophy of Science* 64: S195–212.

McAfee, R. P. and J. McMillan (1987) "Auctions and Bidding," *Journal of Economic Literature* 25: 699–738.

McAfee, R. P. and J. McMillan (1996) "Analyzing the Airwaves Auction," *Journal of Economic Perspectives* 10: 159–75.

McCollum, E. V. and W. Pitz (1917) "The 'Vitamine' Hypothesis and Deficiency Diseases. A Study of Experimental Scurvy," *Journal of Biological Chemistry* 31: 229–53.

McMillan, J. (1994) "Selling Spectrum Rights," *Journal of Economic Perspectives* 8: 145–62.

McMillan, J. (1995) "Why Auction the Spectrum?," *Telecommunications Policy* 19: 191–9.

McMillan, J., M. Rotschild, and R. Wilson (1997) "Introduction," *Journal of Economics and Management Strategy* 6: 425–30.

Mead, W. J., A. Moseidjord, and P. E. Sorensen (1983) "The Rate of Return Earned by Leases Under Cash Bonus Bidding in the OCS Oil and Gas Leases," *Energy Journal* 4: 37–52.

Meyer, B. D., W. K. Viscusi, and D. L. Durbin (1985) "Workers' Compensation and Injury Duration: Evidence from a Natural Experiment," *American Economic Review* 85: 322–40.

Michotte, A. (1946) *La perception de la causalité*. Louvain: Institut Supérieur de Philosophie; Engl. transl. *The Perception of Causality*. London: Methuen, 1963.

Milgrom, P. (1989) "Auctions and Bidding: A Primer," *Journal of Economic Perspectives* 3: 3–22.

Milgrom, P. (1995) "Auctioning the Radio Spectrum," preliminary draft of Milgrom (2004, Ch. 1). http://www.market-design.com/library.html.

Milgrom, P. (1998) "Game Theory and the Spectrum Auctions," *European Economic Review* 42: 771–8.

Milgrom, P. (2000) "Putting Auction Theory to Work: The Simultaneous Ascending Auction," *Journal of Political Economy* 108: 245–72.

Milgrom, P. (2004) *Putting Auction Theory to Work*. Cambridge: Cambridge University Press.

Milgrom, P. R. and R. J. Weber (1982) "A Theory of Auctions and Competitive Bidding," *Econometrica* 50: 1089–122.

Mill, J. S. (1836) "On the Definition of Political Economy and the Method of Investigation Proper to It," in *Collected Works of John Stuart Mill*, Vol. 4. Toronto: University of Toronto Press, 1967, pp. 120–64.

Mill, J. S. (1843) *A System of Logic*. London: Parker.

Miller, D. (2002) "Induction: A Problem Solved," in J. M. Böhm, H. Holweg, and C. Hoock (eds.) *Karl Poppers kritischer Rationalismus heute*. Tübingen: Mohr Siebeck, pp. 81–106.

Miller, R. M. (2002) *Paving Wall Street: Experimental Economics and the Quest for the Perfect Market*. New York: John Wiley & Sons.

Minguzzi, G. F. (1961) "Caratteri espressi e intenzionali dei movimenti: la percezione dell'attesa," *Rivista di psicologia* 55: 157–79.

Mirowski, P. (1989) *More Heat than Light*. Cambridge: Cambridge University Press.

Mirowski, P. (2002) *Machine Dreams*. Cambridge: Cambridge University Press.

Mirowski, P. and E. Nik-Kah (2004) "Markets Made Flesh: Callon, Performativty, and a Crisis in Science Studies, Augmented with Considerations of the FCC Auctions," unpublished paper, University of Notre Dame.

Mongin, P. (1988) "Problèmes de Duhem en théorie de l'utilité espérée," *Fundamenta Scientiae* 9: 299–327.

Mongin, P. (2000) "Les préférences révélées et la formation de la théorie de la demande," *Revue économique* 51: 1125–52.

Mongin, P. (2002) "La conception déductive de l'explication scientifique et l'économie," *Social Science Information* 41:139–65.

Morgan, M. S. (2001) "Models, Stories, and the Economic World," *Journal of Economic Methodology* 8: 361–84.

Morgan, M. S. (2002) "Model Experiments and Models in Experiments," in L. Magnani and N. J. Nersessian (eds.) *Model-Based Reasoning: Science, Technology, Values*. New York: Kluwer, pp. 41–58.

Morgan, M. S. and M. Boumans (2004) "The Secrets Hidden by Two-Dimensionality: Modelling the Economy as a Hydraulic System," in S. de Chadarevian and N. Hopwood (eds.) *Models: The Third Dimension of Science*. Stanford: Stanford University Press, pp. 369–401.

Morrison, M. C. (1998a) "Experiment," in E. Craig (ed.) *The Routledge Encyclopaedia of Philosophy*. London: Routledge, pp. 514–8.

Morrison, M. C. (1998b) "Mediating Models: Between Physics and the Physical World," *Philosophia Naturalis* 35: 65–85.

Morrison, M. C. and M. S. Morgan (1999) "Models as Mediating Instruments," in M. S. Morgan and M. C. Morrison (eds.) *Models as Mediators*. Cambridge: Cambridge University Press, pp. 10–37.

Mulkay, M. and G. N. Gilbert (1986) "Replication and Mere Replication," *Philosophy of the Social Sciences* 16: 21–37.

Murray, J. (2002) *Wireless Nation*. Cambridge, Mass.: Perseus.

Nagel, E. (1961) *The Structure of Science*. New York: Harcourt, Brace & Wold.

Nelson, R. and S. Winter (1982) *An Evolutionary Theory of Economic Change*. Cambridge, Mass.: Harvard University Press.

Newton, I. (1687) *Philosophiae Naturalis Principia Mathematica*. London: Royal Society.

Nik-Kah, E. (unpublished) "Designs on the Mechanism," Ph.D. dissertation, University of Notre Dame.

Nobel Press Release (2002) "The Bank of Sweden Prize in Economic Sciences in Memory of Alfred Nobel" http://www.nobel.se/economics/laureates/2002/.

Norton, J. D. (1996) "Are Thought Experiments Just What You Thought?," *Canadian Journal of Philosophy* 26: 333–66.

Norton, J. D. (2003) "A Material Theory of Induction," *Philosophy of Science* 70: 647–70.

Nye, M. J. (1972) *Molecular Reality*. London: Macdonald.

Offerman, T., J. Sonnemans, and A. Schram (1996) "Value Orientations, Expectations, and Voluntary Contributions in Public Goods," *Economic Journal* 106: 817–45.

Orbell, J., R. Dawes, and A. van de Kragt (1990) "The Limits of Multilateral Promising," *Ethics* 100: 616–27.

Page, T., L. Putterman, and B. Unel (2002) "Voluntary Association in Public Goods Experiments: Reciprocity, Mimicry, and Efficiency," working paper, Brown University.

Palfrey, T. R. and J. E. Prisbey (1996) "Altruism, Reputation and Noise in Linear Public Goods Experiments," *Journal of Public Economics* 61: 409–27.

Palfrey, T. R. and J. E. Prisbey (1997) "Anomalous Behavior in Public Goods Experiments: How Much and Why?," *American Economic Review* 87: 829–46.

Pasteur, L. (1881) "Compte rendu sommaire des expériences rates á Pouilly-le-Fort, prés Melun, sur la vaccination charbonneuse," *Comptes Rendus de l'Academie des Science* 92: 1378–83; Engl. tr. "Summary Report of the Experiments Conducted at Pouilly-le-Fort, Near Melun, on the Anthrax Vaccination," *Yale Journal of Biology and Medicine* 75 (2002): 59–62.

Pasteur, L. (1922) *Oeuvres Complètes*. Paris: Masson.

Paul, J. R. (1971) *A History of Poliomyelitis*. New Haven: Yale University Press.

Pearl, J. (2000) *Causality: Models, Reasoning, and Inference*. Cambridge: Cambridge University Press.

Perrin, J. (1913) *Les atomes*. Paris: Alcan.

Pickering, A. (1995) *The Mangle of Practice: Time, Agency, and Science*. Chicago: University of Chicago Press.

Plott, C. R. (1981) "Experimental Methods in Political Economy: A Tool for Regulatory Research," in A. R. Ferguson (ed.) *Attacking Regulatory Problems*. Cambridge, Mass.: Ballinger, pp. 117–43.

Plott, C. R. (1987) "Dimensions of Parallelism: Some Policy Applications of Experimental Methods," in A. E. Roth (ed.) *Laboratory Experimentation in Economics: Six Points of View*. Cambridge: Cambridge University Press, pp. 193–219.

Plott, C. R. (1991) "Will Economics Become an Experimental Science?," *Southern Economic Journal* 57: 901–19.

Plott, C. R. (1995) "Rational Individual Behaviour in Markets and Social Choice Processes: The Discovered Preference Hypothesis," in K. J. Arrow, E. Colombatto, M. Perlman, and C. Schmidt (eds.) *The Rational Foundations of Economic Behaviour.* London: Macmillan, pp. 225–50.

Plott, C. R. (1996) "Laboratory Experimental Testbeds: Application to the PCS Auction," Social Science Working Paper 957. California Institute of Technology.

Plott, C. R. (1997) "Laboratory Experimental Testbeds: Application to the PCS Auction," *Journal of Economics and Management Strategy* 6: 605–38.

Plott, C. R. (1999) "Policy and the Use of Laboratory Experimental Methodology in Economics," in L. Luini (ed.) *Uncertain Decisions: Bridging Theory and Experiments.* Boston: Kluwer, pp. 293–315.

Plott, C. R. and V. L. Smith, eds. (in press) *The Handbook of Experimental Economics Results.* London: Elsevier.

Pommerehne, W. W., F. Schneider, and P. Zweifel (1982) "Economic Theory of Choice and the Preference Reversal Phenomenon: A Reexamination," *American Economic Review* 72: 569–74.

Popper, K. R. (1934) *Logik der Forschung.* Vienna: Springer; Engl. transl. *Logic of Scientific Discovery.* London: Hutchinson, 1959.

Popper, K. R. (1957) "The Aim of Science," *Ratio* 1: 24–35; reprinted in *Objective Knowledge.* Oxford: Clarendon Press, 1972.

Popper, K. R. (1963) *Conjectures and Refutations.* London: Routledge.

Popper, K. R. (1976) *Unended Quest: An Intellectual Autobiography.* London: Routledge.

Psillos, S. (1999) *Scientific Realism: How Science Tracks the Truth.* London: Routledge.

Putnam, H. (1975) *Philosophical Papers, Vol. 1: Mathematics, Matter and Method,* Cambridge: Cambridge University Press.

Quiggin, J. (1982) "A Theory of Anticipated Utility," *Journal of Economic Behavior and Organization* 3: 323–43.

Quine, W. O. (1953) "Two Dogmas of Empiricism," in *From A Logical Point of View.* Cambridge, Mass.: Harvard University Press, pp. 20–46.

Rabin, M. (1993) "Incorporating Fairness into Game Theory and Economics," *American Economic Review* 83: 1281–302.

Rabin, M. (1998) "Psychology and Economics," *Journal of Economic Literature* 35: 11–46.

Rabin, M. (2002) "A Perspective on Psychology and Economics," *European Economic Review* 46: 657–85.

Radder, H. (1996) *In and About the World: Philosophical Studies of Science and Technology.* Albany: SUNY Press.

Radder, H. (2002) "How Concepts Both Structure the World and Abstract from It," *Review of Metaphysics* 55: 581–613.

Read, D. (in press) "Monetary Incentives, What Are They Good for?," *Journal of Economic Methodology.*

Redhead, M. L. G. (1980) "A Bayesian Reconstruction of the Methodology of Scientific Research Programmes," *Studies in History and Philosophy of Science* 11: 341–7.

Reilly, R. J. (1982) "Preference Reversal: Further Evidence and Some Suggested Modifications in Experimental Design," *American Economic Review* 72: 576–84.

Rice, D. B. and V. L. Smith (1964) "Nature, the Experimental Laboratory, and the Credibility of Hypotheses," *Behavioral Science*; reprinted in V. Smith, *Papers in Experimental Economics*. Cambridge: Cambridge University Press, 1991.

Robbins, L. (1932) *An Essay on the Nature and Significance of Economic Science*. London: Macmillan.

Rosenberg, A. (1992) *Economics: Mathematical Politics or Science of Diminishing Returns?* Chicago: University of Chicago Press.

Rosenberg, A. (1996) "A Field Guide to Recent Species of Naturalism," *British Journal for the Philosophy of Science* 47: 1–29.

Roth, A. E. (1986) "Laboratory Experimentation in Economics," *Economics and Philosophy* 2: 245–73.

Roth, A. E. (1988) "Laboratory Experimentation in Economics: A Methodological Overview," *Economic Journal* 98: 974–1031.

Roth, A. E. (1991) "Game Theory as a Part of Empirical Economics," *Economic Journal* 101: 107–14.

Roth, A. E. (1995) "Introduction to Experimental Economics," in J. H. Kagel and A. E. Roth (eds.) *The Handbook of Experimental Economics*. Princeton: Princeton University Press, pp. 3–109.

Roth, A. E. (2001) "Form and Function in Experimental Design," *Behavioral and Brain Sciences* 24: 427–8.

Roth, A. E. (2002) "The Economist as Engineer: Game Theory, Experimentation, and Computation as Tools for Design Economics," *Econometrica* 70: 1341–78.

Roth, A. E. and M. W. K. Malouf (1979) "Game-Theoretic Models and the Role of Information in Bargaining," *Psychological Review* 86: 574–94.

Roth, A. E. and E. Peranson (1999) "The Redesign of the Matching Market for American Physicians: Some Engineering Aspects of Economic Design," *American Economic Review* 89: 748–80.

Rubinstein, A. (2001) "A Theorist's View of Experiments," *European Economic Review* 45: 615–28.

Safra, Z., U. Segal, and A. Spivak (1990a) "Preference Reversals and Non-expected Utility," *American Economic Review* 80: 922–30.

Safra, Z., U. Segal, and A. Spivak (1990b) "The Becker-DeGroot-Marschak Mechanism and Non-expected Utility: A Testable Approach," *Journal of Risk and Uncertainty* 3: 177–90.

Salanti, A. (1994) "On the Lakatosian Apple of Discord in the History and Methodology of Economics," *Finnish Economic Papers* 7: 30–41.

Salmon, P. (1998) "Free Riding as a Mechanism," in R. E. Backhouse, D. M. Hausman, U. Mäki, and A. Salanti (eds.) *Economics and Methodology: Crossing Boundaries*. London: MacMillan, pp. 62–87.

Salmon, W. C. (1984) *Scientific Explanation and the Causal Structure of the World*. Princeton: Princeton University Press.

Salmon, W. C. (1988) "Rational Prediction," in A. Grünbaum and W. C. Salmon (eds.) *The Limitations of Deductivism*. Berkeley and Los Angeles: University of California Press, pp. 47–60.

Salmon, W. C. (1990) "Rationality and Objectivity in Science, *or* Tom Kuhn Meets Tom Bayes," in C. W. Savage (ed.) *Scientific Theories*. Minnesota Studies in the Philosophy of Science, Vol. 14. Minneapolis: University of Minnesota Press, pp. 175–204.

Samuelson, P. (1938) "A Note on the Pure Theory of Consumer's Behavior," *Economica* 5: 61–71.

Savage, L. J. (1954) *The Foundations of Statistics*. New York: Dover Publications, 2nd ed. 1972.

Sawyer, K. R., C. Beed, and H. Sankey (1997) "Underdetermination in Economics. The Duhem-Quine Thesis," *Economics and Philosophy* 13: 1–23.

Schelling, T. C. (1978) *Micromotives and Macrobehavior*. New York: Norton.

Schotter, A. (1998) "A Practical Person's Guide to Mechanism Selection: Some Lessons from Experimental Economics," in M. Majumdar (ed.) *Organization with Incomplete Information*. Cambridge: Cambridge University Press.

Segal, U. (1988) "Does the Preference Reversals Phenomenon Necessarily Contradict the Independence Axiom?," *American Economic Review* 28: 175–202.

Sen, A. (1973) "Behaviour and the Concept of Preference," *Economica* 40: 241–59.

Sen, A. (1993) "Internal Consistency of Choice," *Econometrica* 61: 495–521.

Siakantaris, N. (2000) "Experimental Economics Under the Microscope," *Cambridge Journal of Economics* 24: 267–81.

Simon, H. A. (1969) *The Sciences of the Artificial*. Boston: MIT Press.

Slovic, P. (1995) "The Construction of Preferences," *American Psychologist* 50: 364–71.

Smith, V. L. (1962) "An Experimental Study of Competitive Market Behavior," *Journal of Political Economy* 70: 111–37.

Smith, V. L. (1976) "Experimental Economics: Induced Value Theory," *American Economic Review* 66: 274–7.

Smith, V. L. (1982) "Microeconomic Systems as an Experimental Science," *American Economic Review* 72: 923–55.

Smith, V. L. (1989) "Theory, Experiment and Economics," *Journal of Economic Perspectives* 3: 151–69.

Smith, V. L. (1991a) *Papers in Experimental Economics*. Cambridge: Cambridge University Press.

Smith, V. L. (1991b) "Rational Choice: The Contrast Between Economics and Psychology," *Journal of Political Economy* 99: 877–97.

Smith, V. L. (1992) "Game Theory and Experimental Economics: Beginnings and Early Influences," in E. R. Weintraub (ed.) *Towards A History of Game Theory*. History of Political Economy Supplement, Vol. 24. Durham: Duke University Press, pp. 241–82 .

Smith, V. L. (1994) "Economics in the Laboratory," *Journal of Economic Perspectives* 8: 113–31.

Smith, V. L. (2002) "Method in Experiment: Rhetoric and Reality," *Experimental Economics* 5: 91–110.

Smith, V. L. (in press) "Experimental Methods in (Neuro)Economics," in *Encyclopedia of Cognitive Science*.

Smith, V. L. and J. M. Walker (1993) "Monetary Rewards and Decision Costs in Experimental Economics," *Economic Inquiry* 31: 245–61.

Sober, E. (1988) *Reconstructing the Past: Parsimony, Evolution, and Inference*. Cambridge, Mass.: MIT Press.

Sober, E. (2002) "Bayesianism – Its Scope and Limits," in R. Swinburne (ed.) *Bayes' Theorem*. Proceedings of the British Academy Press, Vol. 113: 21–38.

Soberg, M. (in press) "The Duhem-Quine Thesis and Experimental Economics: A Reinterpretation," *Journal of Economic Methodology*.

Sorensen, R. (1992) *Thought Experiments*. Oxford: Oxford University Press.

Starmer, C. (1999) "Experiments in Economics . . . (Should We Trust the Dismal Scientists in White Coats?)," *Journal of Economic Methodology* 6: 1–30.

Starmer, C. (in press) "On Testing Game Theory," *Journal of Economic Methodology*.

Starmer, C. and R. Sugden (1991) "Does the Random-Lottery Incentive System Elicit True Preferences? An Experimental Investigation," *American Economic Review* 81: 971–8.

Strand, R., R. Fjelland, and T. Flatmark (1996) "In Vivo Interpretation of In Vitro Effect Studies," *Acta Biotheoretica* 44: 1–21.

Sugden, R. (1984) "Reciprocity: The Supply of Public Goods through Voluntary Contributions," *Economic Journal* 94: 772–87.

Sugden, R. (2000) "Credible Worlds: The Status of Theoretical Models in Economics," *Journal of Economic Methodology* 7: 1–31.

Sugden, R. (in press) "Experiments as Exhibits and Experiments as Tests," *Journal of Economic Methodology*.

Suppe, F., ed. (1977) *The Structure of Scientific Theories*. Urbana: University of Illinois Press.

Suppe, F. (1989) *The Semantic Conception of Theories and Scientific Realism*. Urbana: University of Illinois Press.

Suppes, P. (1984) *Probabilistic Metaphysics*. London: Blackwell.

Tammi, T. (1999) "Incentives and Preference Reversals: Escape Moves and Community Decisions," *Journal of Economic Methodology* 6: 351–80.

Thagard, P. (1999) *How Scientists Explain Disease*. Princeton: Princeton University Press.

Thaler, R. H. (1988) "Anomalies: The Winner's Curse," *Journal of Economic Perspectives* 2: 191–202.

Thaler, R. H. and A. Tversky (1990) "Anomalies: Preference Reversals," *Journal of Economic Perspectives* 4: 201–11.

Titmuss, R. M. (1970) *The Gift Relationship: From Human Blood to Social Policy*. London: Allen & Unwin.

Tversky, A., P. Slovic, and D. Kahneman (1990) "The Causes of Preference Reversals," *American Economic Review* 80: 204–17.

Uebel, T., ed. (1991) *Rediscovering the Forgotten Vienna Circle*. Dordrecht: Kluwer.

van Fraassen, B. (1980) *The Scientific Image*. Oxford: Oxford University Press.

van Vugt, M. (2001) "Self-Interest as Self-Fulfilling Prophecy," *Behavioral and Brain Sciences* 24: 429–30.

Vickrey, W. (1961) "Counterspeculation, Auctions, and Competitive Sealed Tenders," *Journal of Finance* 16: 8–37.

von Neumann, J. and O. Morgenstern (1944) *The Theory of Games and Economic Behavior*. Princeton: Princeton University Press.

Watkins, J. (1984) *Science and Scepticism*. Princeton: Princeton University Press.

Weber, R. J. (1997) "Making More from Less: Strategic Demand Reduction in the FCC Spectrum Auctions," *Journal of Economics and Management Strategy* 6: 529–48.

Weibull, J. W. (2002) "Testing Game Theory," SSE Discussion Paper 382, Stockholm School of Economics.

Wilde, L. L. (1981) "On the Use of Laboratory Experiments in Economics," in J. C. Pitt (ed.) *Philosophy in Economics*. Dordrecht: Reidel, pp. 137–48.

Wilson, R. B. (1977) "A Bidding Model of Perfect Competition," *Review of Economic Studies* 44: 511–8.

Wilson, R. B. (2002) "Architecture of Power Markets," *Econometrica* 70: 1299–340.

Woodward, J. (1989) "Data and Phenomena," *Synthèse* 79: 393–472.

Woodward, J. (2002) "Experimentation, Causal Inference, and Instrumental Realism," in H. Radder (ed.) *The Philosophy of Scientific Experimentation*. Pittsburgh: Pittsburgh University Press, pp. 87–118.

Woodward, J. (2003) *Making Things Happen: A Theory of Causal Explanation*. Oxford: Oxford University Press.

Worrall, J. (1978) "The Ways in which the Methodology of Scientific Research Programmes Improves on Popper's Methodology," in G. Andersson and A. Radnitzky (eds.) *Progress and Rationality of Science*. Dordrecht: Reidel, pp. 45–70.

Worrall, J. (1985) "Scientific Discovery and Theory-Confirmation," in J. Pitt (ed.) *Change and Progress in Modern Science*. Dordrecht: Reidel, pp. 301–31.

Worrall, J. (1989) "Why Both Popper and Watkins Fail to Solve the Problem of Induction," in F. D'Agostino and I. C. Jarvie (eds.) *Freedom and Rationality: Essays in Honour of John Watkins*. Dordrecht: Kluwer, pp. 257–96.

Worrall, J. (1993) "Falsification, Rationality, and the Duhem Problem," in J. Earman, A. Janis, G. Massey, and N. Rescher (eds.) *Philosophical Problems of the Internal and External Worlds: Essays on the Philosophy of Adolf Grünbaum*. Pittsburgh: University of Pittsburgh Press, pp. 329–70.

Yaari, M. E. (1987) "The Dual Theory of Choice Under Risk," *Econometrica* 55: 95–115.

Zahar, E. (1976) "Why Did Einstein's Programme Supersede Lorentz's," in C. Howson (ed.) *Method and Appraisal in the Physical Sciences*. Cambridge: Cambridge University Press, pp. 211–75.

Zahar, E. (1983) "Logic of Discovery or Psychology of Invention?," *British Journal for the Philosophy of Science* 34: 243–61.

Index

randomization, 63, 64, 78–79
  background factors in, 134
  experimental control with, 78–80
  experimental design with, 66
  external validity with, 144
  other factors problem with, 132–135
rational belief, 114
rational choice theory, 236
  independence principle and, 100
rationality principle, preference reversals
  with, 93
realism, 87, 90
  experimental economics and, 158, 221n
real-world system. *See* target system
reciprocators, 23–25
recruiting, 33–34
reduction principle, 101
reductionism, 154
  causation and, 72
  economics and, 154
refutation
  hypothesis testing with, 50–52, 58, 59
  model, 220
repetitions, 14, 152, 228
replications, 13–15, 68
  failed, 37–38
  phenomena, 68
  purpose of, 15
  repetitions vs., 14
  value of, 14
representative sampling, 197
research program, 120, 148
reservation price, BDM with, 97
restrictive assumptions, 166
results, 35–38
  qualitative, 36
  quantitative, 36
Rice, Donald, 253–254
ripple tank model, 214
RLS. *See* Random Lottery Selection
RNNE. *See* non-cooperative equilibrium
  with risk-neutral bidders
Robbins, Lionel, 2
robustness, 174
  external validity vs., 224–229
  model with, 225
  of phenomena, 222–224
  of preference reversals phenomenon,
  226, 229
  testbed experiments and, 172
Roth, A. E., 40, 52, 182

Rubinstein, Ariel, 89, 107
rule testing, economic engineering, 175–178

Safra, Zvi, 98, 100, 101, 103–104, 112, 121,
  123–125, 250–252
  hypothesis of, 107
  model of, 105
saliency, monetary incentives with, 232
Samuelson, Paul, 3, 91
Saunders, S., 8
Schelling, Thomas, 205–206
Schram, A., 24
science
  applied, 184, 187
  community, 132
  continuity of, 17
  fact in, 40
  hypotheses, 52
  natural, 5–6
  normal, 16, 156
  philosophy of, 4–6, 53
  revolutions in, 16
scientific genius, myth of, 17
scientific growth model, 200–202
scientific method, 4–5, 6, 49, 149
  evidence in, 40
  hypothesis testing with, 39
  models of, 39
  rhetoric of, 7
scientific reasoning, 39
scientific theories, theory of, 53
screening conditions, 71–78
screening device, 148
Searching for facts, 40
Segal, Uzi, 101, 121, 123
Selten, Rheinhard, 1
semantic theorists, 210
Sen, Amartya, 239
Shanks, Niall, 195
Simon, Herbert, 1, 213
simulations
  definition of, 213
  external validity inference for, 215
  knowledge producing device as, 216
  model, experiment with, 213–217
  simultaneous ascending-bid auction, 167,
  176
  economic engineering design of, 181
  external validity checking for, 179
Slovic, Paul, 92, 95, 97, 98, 104, 113, 122,
  126–127, 225, 242

Printed in the United States
By Bookmasters